Automation for Robotics

Series Editor
Hisham Abou Kandil

Automation for Robotics

Luc Jaulin

WILEY

First published 2015 in Great Britain and the United States by ISTE Ltd and John Wiley & Sons, Inc.

Apart from any fair dealing for the purposes of research or private study, or criticism or review, as permitted under the Copyright, Designs and Patents Act 1988, this publication may only be reproduced, stored or transmitted, in any form or by any means, with the prior permission in writing of the publishers, or in the case of reprographic reproduction in accordance with the terms and licenses issued by the CLA. Enquiries concerning reproduction outside these terms should be sent to the publishers at the undermentioned address:

ISTE Ltd
27-37 St George's Road
London SW19 4EU
UK

www.iste.co.uk

John Wiley & Sons, Inc.
111 River Street
Hoboken, NJ 07030
USA

www.wiley.com

© ISTE Ltd 2015

The rights of Luc Jaulin to be identified as the author of this work have been asserted by him in accordance with the Copyright, Designs and Patents Act 1988.

Library of Congress Control Number: 2014955868

British Library Cataloguing-in-Publication Data
A CIP record for this book is available from the British Library
ISBN 978-1-84821-798-0

Contents

INTRODUCTION	vii
CHAPTER 1. MODELING	1
1.1. Linear systems	1
1.2. Mechanical systems	2
1.3. Servomotors	4
1.4. Exercises	4
1.5. Solutions	21
CHAPTER 2. SIMULATION	47
2.1. Concept of vector field	47
2.2. Graphical representation	49
2.2.1. Patterns	50
2.2.2. Rotation matrix	50
2.2.3. Homogeneous coordinates	52
2.3. Simulation	54
2.3.1. Euler's method	54
2.3.2. Runge–Kutta method	55
2.3.3. Taylor's method	56
2.4. Exercises	56
2.5. Solutions	67

CHAPTER 3. LINEAR SYSTEMS 85

3.1. Stability 85
3.2. Laplace transform 87
   3.2.1. Laplace variable 87
   3.2.2. Transfer function 88
   3.2.3. Laplace transform 88
   3.2.4. Input–output relation 90
3.3. Relationship between state and transfer representations 90
3.4. Exercises 92
3.5. Solutions 103

CHAPTER 4. LINEAR CONTROL 127

4.1. Controllability and observability 128
4.2. State feedback control 129
4.3. Output feedback control 130
4.4. Summary 133
4.5. Exercises 134
4.6. Solutions 150

CHAPTER 5. LINEARIZED CONTROL 185

5.1. Linearization 185
   5.1.1. Linearization of a function 185
   5.1.2. Linearization of a dynamic system 187
   5.1.3. Linearization around an operating point . . . 187
5.2. Stabilization of a nonlinear system 188
5.3. Exercises 191
5.4. Solutions 207

BIBLIOGRAPHY 235

INDEX 237

Introduction

I.1. State representation

Biological, economic and other mechanical systems surrounding us can often be described by a differential equation such as:

$$\begin{cases} \dot{\mathbf{x}}(t) = \mathbf{f}(\mathbf{x}(t), \mathbf{u}(t)) \\ \mathbf{y}(t) = \mathbf{g}(\mathbf{x}(t), \mathbf{u}(t)) \end{cases}$$

under the hypothesis that the time t in which the system evolves is continuous [JAU 05]. The vector $\mathbf{u}(t)$ is the *input* (or *control*) of the system. Its value may be chosen arbitrarily for all t. The vector $\mathbf{y}(t)$ is the *output* of the system and can be measured with a certain degree of accuracy. The vector $\mathbf{x}(t)$ is called the *state* of the system. It represents the memory of the system, in other words the information needed by the system in order to predict its own future, for a known input $\mathbf{u}(t)$. The first of the two equations is called the *evolution equation*. It is a differential equation that enables us to know where the state $\mathbf{x}(t)$ is headed knowing its value at the present moment t and the control $\mathbf{u}(t)$ that we are currently exerting. The second equation is called the *observation equation*. It allows us to calculate the output vector $\mathbf{y}(t)$, knowing the state and control at time t. Note, however, that, unlike the evolution equation, this equation is not a differential equation as it does not involve the derivatives of the signals. The two equations given above form the *state representation* of the system.

It is sometimes useful to consider a discrete time k, with $k \in \mathbb{Z}$, where $\mathbb{Z}$ is the set of integers. If, for instance, the universe is being considered as a computer, it is possible to consider that the time k is discrete and synchronized to the clock of the microprocessor. Discrete-time systems often respect a recurrence equation such as:

$$\begin{cases} \mathbf{x}(k+1) = \mathbf{f}(\mathbf{x}(k), \mathbf{u}(k)) \\ \mathbf{y}(k) = \mathbf{g}(\mathbf{x}(k), \mathbf{u}(k)) \end{cases}$$

The first objective of this book is to understand the concept of state representation through numerous exercises. For this, we will consider, in Chapter 1, a large number of varied exercises and show how to reach a state representation. We will then show, in Chapter 2, how to simulate a given system on a computer using its state representation.

The second objective of this book is to propose methods to *control* the systems described by state equations. In other words, we will attempt to build *automatic* machines (in which humans are practically not involved, except to give orders, or *setpoints*), called *controllers* capable of *domesticating* (changing the behavior in a desired direction) the systems being considered. For this, the controller will have to compute the inputs $\mathbf{u}(t)$ to be applied to the system from the (more or less noisy) knowledge of the outputs $\mathbf{y}(t)$ and from the setpoints $\mathbf{w}(t)$ (see Figure I.1).

From the point of view of the user, the system, referred to as a *closed-loop system*, with input $\mathbf{w}(t)$ and output $\mathbf{y}(t)$, will have a suitable behavior. We will say that we have *controlled* the system. With this objective of control, we will, in a first phase, only look at linear systems, in other words when the functions f and g are assumed linear. Thus, in the continuous-time case, the state equations of the system are written as:

$$\begin{cases} \dot{\mathbf{x}}(t) = \mathbf{A}\mathbf{x}(t) + \mathbf{B}\mathbf{u}(t) \\ \mathbf{y}(t) = \mathbf{C}\mathbf{x}(t) + \mathbf{D}\mathbf{u}(t) \end{cases}$$

and in the discrete-time case, they become:

$$\begin{cases} \mathbf{x}(k+1) = \mathbf{A}\mathbf{x}(k) + \mathbf{B}\mathbf{u}(k) \\ \mathbf{y}(k) = \mathbf{C}\mathbf{x}(k) + \mathbf{D}\mathbf{u}(k) \end{cases}$$

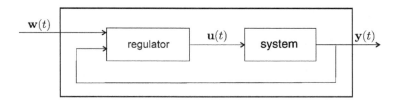

Figure I.1. *Closed loop concept illustrating the control of a system*

The matrices $\mathbf{A}, \mathbf{B}, \mathbf{C}, \mathbf{D}$ are called *evolution, control, observation* and *direct matrices*. A detailed analysis of these systems will be performed in Chapter 3. We will then explain, in Chapter 4, how to stabilize these systems. Finally, we will show in Chapter 5 that around certain points, called *operating points*, nonlinear systems behave like linear systems. It will then be possible to stabilize them using the same methods as those developed for the linear case.

Finally, this book is accompanied by numerous MATLAB programs available at:

http//www.ensta-bretagne.fr/jaulin/isteauto.html

I.2. Exercises

EXERCISE I.1.– Underwater robot

The underwater robot *Saucisse* of the Superior National School of Advanced Techniques (SNSAT) Bretagne [JAU 09], whose photo is given in Figure I.2, is a control system. It includes a computer, three propellers, a camera, a compass and a sonar. What does the input vector u, the output vector y, the state vector x and the setpoint w correspond to in this context? Where does the computer come in the control loop?

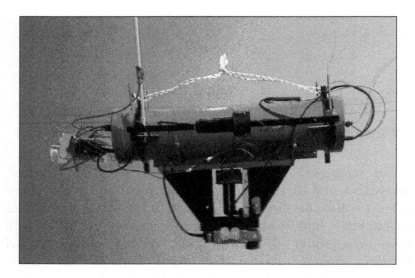

Figure I.2. *Controlled underwater robot*

EXERCISE I.2.– Sailing robot

The sailing robot *Vaimos* (French Research Institute for Exploitation of the Sea (FRIES) and SNSAT Bretagne) in Figure I.3 is also a control system [JAU 12a, JAU 12b]. It is capable of following paths by itself, such as the one drawn in Figure I.3. It has a rudder and a sail adjustable using a sheet. It also has an anemometer on top of the mast, a compass and a Global Positioning System (GPS). Describe what the input vector u, the output vector y, the state vector x and the setpoint w may correspond to.

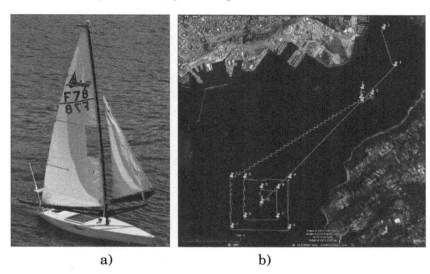

a)   b)

Figure I.3. *Sailing robot Vaimos a) and a path followed by Vaimos b). The zig-zags in the path are due to Vaimos having to tack in order to sail against the wind*

I.3. Solutions

Solution to Exercise I.1 (underwater robot)

The input vector $u \in \mathbb{R}^3$ corresponds to the electric voltage given to the three propellers and the output vector $y(t)$ includes the compass, the sonar data and the images taken by the cameras. The state vector x corresponds to the position, orientation and speeds of the robot. The setpoint w is requested by the supervisor. For instance, if we want to perform a course control, the setpoint w will be the desired speed and course for the robot. The controller is a pogram executed by the computer.

Solution to Exercise I.2 (sailing robot)

The input vector $u \in \mathbb{R}^2$ corresponds to the length of the sail sheet $\delta_v^{\max}$ and to the angle of the rudder δ_g. The output vector $y \in \mathbb{R}^4$ includes the GPS data m, the ultrasound anemometer (weather vane on top of the mast) ψ and the compass θ. The setpoint w indicates here the segment ab to follow. Figure I.4 illustrates this control loop. A supervisor, not represented on the figure, takes care of sequencing the segments to follow in such a way that the robot follows the desired path (here 12 segments forming a square box followed by a return to port).

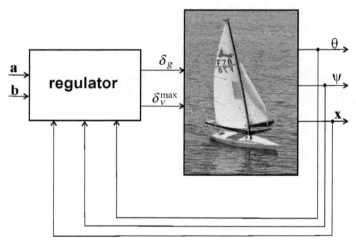

Figure I.4. *Control loop of the sailing robot*

1
Modeling

We will call *modeling* the step that consists of finding a more or less accurate state representation of the system we are looking at. In general, constant parameters appear in the state equations (such as the mass or the inertial moment of a body, the coefficient of viscous friction, the capacitance of a capacitor, etc.). In these cases, an identification step may prove to be necessary. In this book, we will assume that all the parameters are known, otherwise we invite the reader to consult Eric Walter's book [WAL 14] for a broad range of identification methods. Of course, no systematic methodology exists that can be used to model a system. The goal of this chapter and of the following exercises is to present, using several varied examples, how to obtain a state representation.

1.1. Linear systems

In the continuous-time case, linear systems can be described by the following state equations:

$$\begin{cases} \dot{\mathbf{x}}(t) = \mathbf{A}\mathbf{x}(t) + \mathbf{B}\mathbf{u}(t) \\ \mathbf{y}(t) = \mathbf{C}\mathbf{x}(t) + \mathbf{D}\mathbf{u}(t) \end{cases}$$

Linear systems are rather rare in nature. However, they are relatively easy to manipulate using linear algebra techniques and often approximate in an acceptable manner the nonlinear systems around their operating point.

1.2. Mechanical systems

The fundamental principle of dynamics allows us to easily find the state equations of mechanical systems (such as robots). The resulting calculations are relatively complicated for complex systems and the use of computer algebra systems may prove to be useful. In order to obtain the state equations of a mechanical system composed of several subsystems $\mathcal{S}_1, \mathcal{S}_2, \ldots, \mathcal{S}_m$, assumed to be rigid, we follow three steps:

1) *Obtaining the differential equations.* For each subsystem $\mathcal{S}_k$, with mass m and inertial matrix $\mathbf{J}$, the following relations must be applied:

$$\sum_i \mathbf{f}_i = m\mathbf{a}$$
$$\sum_i \mathcal{M}_{\mathbf{f}_i} = \mathbf{J}\dot{\omega}$$

where the $\mathbf{f}_i$ are the forces acting on the subsystem $\mathcal{S}_k$, $\mathcal{M}_{\mathbf{f}_i}$ represents the torque created by the force $\mathbf{f}_i$ on $\mathcal{S}_k$, with respect to its center. The vector a represents the tangential acceleration of $\mathcal{S}_k$ and the vector $\dot{\omega}$ represents the angular acceleration of $\mathcal{S}_k$. After decomposing these $2m$ vectorial equations according to their components, we obtain $6m$ scalar differential equations such that some of them might be degenerate.

2) *Removing the components of the internal forces.* In differential equations there are the so-called *bonding* forces, which are internal to the whole mechanical system, even though they are external to each subsystem composing it. They represent the action of a subsystem $\mathcal{S}_k$ on another subsystem $\mathcal{S}_\ell$. Following the action–reaction principle, the

existence of such a force, denoted by $f^{k,\ell}$, implies the existence of another force $f^{\ell,k}$, representing the action of $\mathcal{S}_\ell$ on $\mathcal{S}_k$, such that $f^{\ell,k} = -f^{k,\ell}$. Through a formal manipulation of the differential equations and by taking into account the equations due to the action-reaction principle, it is possible to remove the internal forces. The resulting number of differential equations has to be reduced to the number n of degrees of freedom $q_1, \ldots, q_n$ of the system.

3) *Obtaining the state equations*. We then have to isolate the second derivative $\ddot{q}_1, \ldots, \ddot{q}_n$ from the set of n differential equations in such a manner to obtain a vectorial relation such as:

$$\ddot{\mathbf{q}} = \mathbf{f}(\mathbf{q}, \dot{\mathbf{q}}, \mathbf{u})$$

where u is the vector of external forces that are not derived from a potential (in other words, those which we apply to the system). The state equations are then written as:

$$\frac{d}{dt}\begin{pmatrix} \mathbf{q} \\ \dot{\mathbf{q}} \end{pmatrix} = \begin{pmatrix} \dot{\mathbf{q}} \\ \mathbf{f}(\mathbf{q}, \dot{\mathbf{q}}, \mathbf{u}) \end{pmatrix}$$

A mechanical system whose dynamics can be described by the relation $\ddot{\mathbf{q}} = \mathbf{f}(\mathbf{q}, \dot{\mathbf{q}}, \mathbf{u})$ will be referred to as *holonomic*. For a holonomic system, q and $\dot{\mathbf{q}}$ are thus independent. If there is a so-called *non-holonomic* constraint that links the two of them (of the form $h(\mathbf{q}, \dot{\mathbf{q}}) = 0$), the system will be referred to as *non-holonomic*. Such systems may be found for instance in mobile robots with wheels [LAU 01]. Readers interested in more details on the modeling of mechanical systems may consult [KHA 07].

1.3. Servomotors

A mechanical system is controlled by forces or torques and obeys a dynamic model that depends on many poorly known coefficients. This same mechanical system represented by a kinematic model is controlled by positions, velocities or accelerations. The kinematic model depends on well-known geometric coefficients and is a lot easier to put into equations. In practice, we move from a dynamic model to its kinematic equivalent by adding servomotors. In summary, a servomotor is a direct current motor with an electrical control circuit and a sensor (of the position, velocity or acceleration). The control circuit computes the voltage u to give to the motor in order for the value measured by the sensor corresponds to the setpoint w. In practice, the signal w is generally given in the form of a square wave called *pulse-width modulation* (PWM)). There are three types of servomotors:

– the *position servo*. The sensor measures the position (or the angle) x of the motor and the control rule is expressed as $u = k(x - w)$. If k is large, we may conclude that $x \simeq w$;

– the *velocity servo*. The sensor measures the velocity (or the angular velocity) $\dot{x}$ of the motor and the control rule is expressed as $u = k(\dot{x} - w)$. If k is large, we have $\dot{x} \simeq w$;

– the *acceleration servo*. The sensor measures the acceleration (tangential or angular) $\ddot{x}$ of the motor and the control rule is expressed as $u = k(\ddot{x} - w)$. If k is large, we have $\ddot{x} \simeq w$.

1.4. Exercises

EXERCISE 1.1.– Integrator

The integrator is a linear system described by the differential equation $\dot{y} = u$. Find a state representation for this system. Give this representation in a matrix form.

EXERCISE 1.2.– Second order system

Let us consider the system with input u and output y described by the second order differential equation:

$$\ddot{y} + a_1 \dot{y} + a_0 y = bu$$

Taking $\mathbf{x} = (y, \dot{y})$, find a state equation for this system. Give it in matrix form.

EXERCISE 1.3.– Mass-spring system

Let us consider a system with input u and output q_1 as shown in Figure 1.1 (u is the force applied to the second carriage, q_i is the deviation of the ith carriage with respect to its point of equilibrium, k_i is the stiffness of the ith spring and α is the coefficient of viscous friction).

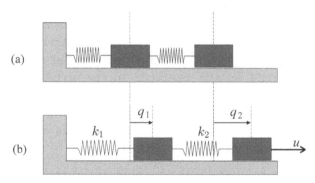

Figure 1.1. *a) Mass-spring system at rest, b) system in any state*

Let us take the state vector:

$$\mathbf{x} = (q_1, q_2, \dot{q}_1, \dot{q}_2)^{\mathrm{T}}$$

1) Find the state equations of the system.

2) Is this system linear?

EXERCISE 1.4.– Simple pendulum

Let us consider the pendulum in Figure 1.2. The input of this system is the momentum u exerted on the pendulum around its axis. The output is $y(t)$, the algebraic distance between the mass m and the vertical axis:

1) Determine the state equations of this system.

2) Express the mechanical energy E_m as a function of the state of the system. Show that the latter remains constant when the momentum u is nil.

EXERCISE 1.5.– Dynamic modeling of an inverted rod pendulum

Let us consider the so-called *inverted rod pendulum* system, which is composed of a pendulum of length ℓ placed in an unstable equilibrium on a carriage, as represented in Figure 1.3. The value u is the force exerted on the carriage of mass M, x indicates the position of the carriage, θ is the angle between the pendulum and the vertical axis and $\vec{R}$ is the force exerted by the carriage on the pendulum. At the extremity B of the pendulum, a point mass m is fixated. We may ignore the mass of the rod. Finally, A is the point of articulation between the rod and the carriage and $\vec{\Omega} = \dot{\theta}\vec{k}$ is the rotation vector associated with the rod.

1) Write the fundamental principle of dynamics as applied on the carriage and the pendulum.

2) Show that the velocity vector at point B is expressed by the relation $\mathbf{v}_B = \left(\dot{x} - \dot{\theta}\cos\theta\right)\vec{i} - \dot{\theta}\sin\theta\vec{j}$. Calculate the acceleration $\dot{\mathbf{v}}_B$ of point B.

3) In order to model the inverted pendulum, we will take the state vector $\mathbf{x} = \left(x, \theta, \dot{x}, \dot{\theta}\right)$. Justify this choice.

4) Find the state equations for the inverted rod pendulum.

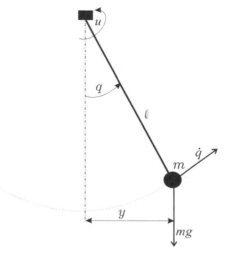

Figure 1.2. *Simple pendulum with state vector* $\mathbf{x} = (q, \dot{q})$

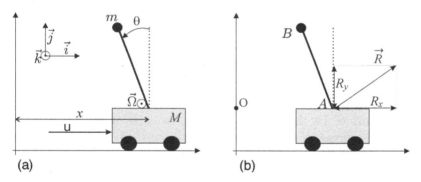

Figure 1.3. *Inverted rod pendulum*

EXERCISE 1.6.– Kinematic modeling of an inverted rod pendulum

In a kinematic model, the inputs are no longer forces or moments, but kinematic variables, in other words positions, velocities or accelerations. It is the role of servomotors to translate these kinematic variables into forces or moments.

Let us take the state equations of the inverted rod pendulum established in the previous exercise:

$$\frac{d}{dt}\begin{pmatrix} x \\ \theta \\ \dot{x} \\ \dot{\theta} \end{pmatrix} = \begin{pmatrix} \dot{x} \\ \dot{\theta} \\ \frac{-m\sin\theta(\ell\dot{\theta}^2 - g\cos\theta)}{M+m\sin^2\theta} \\ \frac{\sin\theta((M+m)g - m\ell\dot{\theta}^2\cos\theta)}{\ell(M+m\sin^2\theta)} \end{pmatrix} + \begin{pmatrix} 0 \\ 0 \\ \frac{1}{M+m\sin^2\theta} \\ \frac{\cos\theta}{\ell(M+m\sin^2\theta)} \end{pmatrix} u$$

1) Instead of taking the force u on the carriage as input, let us rather take the acceleration $a = \ddot{x}$. What does the state model become?

2) Show how, by using a proportional control such as $u = K(a - \ddot{x})$ with large K, it is possible to move from a dynamic model to a kinematic model. In what way does this control recall the servomotor principle or the operational amplifier principle?

EXERCISE 1.7.– Segway

The segway represented on the left side of Figure 1.4 is a vehicle with two wheels and a single axle. It is stable since it is controlled. In the modeling step, we will of course assume that the engine is not controlled.

Its open loop behavior is very close to that of the planar unicycle represented in Figure 1.4 on the right hand side. In this figure, u represents the exerted momentum between the body and the wheel.

The link between these two elements is a pivoting pin. We will denote by B the center of gravity of the body and by A that of the wheel. C is a fix point on the disk. Let us denote by α the angle between the vector $\overrightarrow{AC}$ and the horizontal axis and by θ the angle between the body of the unicycle and the vertical axis. This system has two degrees of freedom α and θ. The state of out system is given by the vector $\mathbf{x} = (\alpha, \theta, \dot{\alpha}, \dot{\theta})^{\mathrm{T}}$.

The parameters of our system are:

– for the disk: its mass M, its radius a, its moment of inertia J_M;

– for the pendulum: its mass m, its moment of inertia J_p, the distance ℓ between its center of gravity and the center of the disk.

Find the equations of the systems.

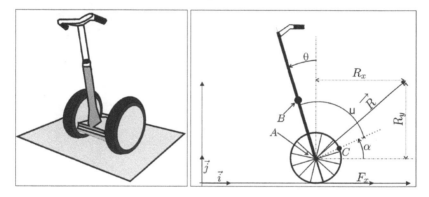

Figure 1.4. *The segway has two wheels and one axle*

EXERCISE 1.8.– Hamilton's method

Hamilton's method allows us to obtain the state equations of a conservative mechanical system (in other words, whose energy is conserved) only from the expression of a single function: its energy. For this, we define the *Hamiltonian* as the mechanical energy of the system, in other words the sum of the potential energy and the kinetic energy. The Hamiltonian can be expressed as a function $H(\mathbf{q}, \mathbf{p})$ of the degrees of freedom $\mathbf{q}$ and of the associated amount of movement (or kinetic moments in the case of a rotation) $\mathbf{p}$. The Hamilton equations are written as:

$$\begin{cases} \dot{\mathbf{q}} = \frac{\partial H(\mathbf{q},\mathbf{p})}{\partial \mathbf{p}} \\ \dot{\mathbf{p}} = -\frac{\partial H(\mathbf{q},\mathbf{p})}{\partial \mathbf{q}} \end{cases}$$

1) Let us consider the simple pendulum shown in Figure 1.6. This pendulum has a length of ℓ and is composed of a single point mass m. Calculate the Hamiltonian of the system. Deduce the state equations from this.

2) Show that if a system is described by Hamilton equations, then the Hamiltonian is constant.

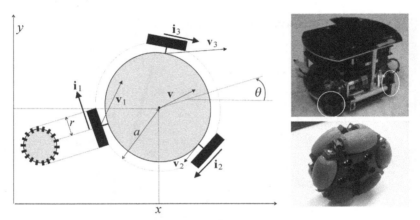

Figure 1.5. *Holonomic robot with omni wheels*

EXERCISE 1.9.– Omnidirectional robot

Let us consider the robot with three omni wheels, as shown in Figure 1.5. An omni wheel is a wheel equipped with a multitude of small rollers over its entire periphery that allow it to slide sideways (in other words perpendicularly to its nominal movement direction). Let us denote by $\mathbf{v}_i$ the velocity vector of the contact point of the ith wheel. If $\mathbf{i}_i$ is the normed direction vector indicating the nominal movement direction of the wheel, then the component of $\mathbf{v}_i$ according to $\mathbf{i}_i$ corresponds to the rotation ω_i of the wheel whereas its complementary component (perpendicular to $\mathbf{i}_i$) is linked to the rotation of the peripheral rollers. If r is the radius of the wheel, then we have the relation $r\omega_i = \langle \mathbf{v}_i, \mathbf{i}_i \rangle = \|\mathbf{v}_i\| \cdot \|\mathbf{i}_i\| \cdot \cos\alpha_i$, where $\alpha_i = \widehat{\cos(\mathbf{v}_i, \mathbf{i}_i)}$. If $\cos\alpha_i = \pm 1$, the wheel is in its nominal state, i.e. it behaves like a classical wheel. If $\cos \alpha_i = 0$, the wheel no longer turns and it is in a state of skid.

1) Give the state equations of the system. We will use the state vector $\mathbf{x} = (x, y, \theta)$ and the input vector $\omega = (\omega_1, \omega_2, \omega_3)$.

2) Propose a loop that allows us to obtain a model tank described by the following state equations:

$$\begin{cases} \dot{x} = v \cos \theta \\ \dot{y} = v \sin \theta \\ \dot{\theta} = u_1 \\ \dot{v} = u_2 \end{cases}$$

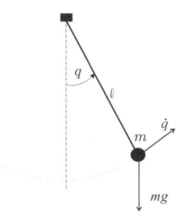

Figure 1.6. *Simple pendulum*

EXERCISE 1.10.– Modeling a tank

The robot tank in Figure 1.7 is composed of two parallel motorized crawlers (or wheels) whose accelerations (which form the inputs u_1 and u_2 of the system) are controlled by two independent motors. In the case where wheels are considered, the stability of the system is ensured by one or two idlers, not represented on the figure. The degrees of freedom of the robot are the x, y coordinates of the center of the axle and its orientation θ.

1) Why can't we choose as state vector the vector $\mathbf{x} = (x, y, \theta, \dot{x}, \dot{y}, \dot{\theta})^{\mathrm{T}}$?

2) Let us denote by v_1 and v_2 the center velocities of each of the motorized wheels. Let us choose as state vector the vector $\mathbf{x} = (x, y, \theta, v_1, v_2)^{\mathrm{T}}$. What might justify such a choice? Give the state equations of the system.

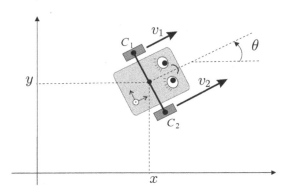

Figure 1.7. *Robot tank viewed from above*

EXERCISE 1.11.– Modeling a car

Let us consider the car as shown in Figure 1.8. The driver of the car (on the left hand side on the figure) has two controls: acceleration of the front wheels (assumed to be motorized) and rotation velocity of the steering wheel. The brakes here represent a negative acceleration. We will denote by δ the angle between the front wheels and the axis of the car, by θ the angle made by the car with respect to the horizontal axis and by (x, y) the coordinates of the middle of the rear axle. The state variables of our system are composed of:

– the position coordinates, in other words all the knowledge necessary to draw the car, more specifically the x, y

coordinates of the center of the rear axle, the orientation θ of the car, and the angle δ of the front wheels;

– the kinetic coordinate v representing the velocity of the center of the front axle (indeed, the sole knowledge of this value and the position coordinates allows to calculate all the velocities of all the other elements of the car).

Calculate the state equations of the system. We will assume that the two wheels have the same velocity v (even though in reality, the inner wheel during a turn is slower than the outer one). Thus, as illustrated on the right-hand side figure, everything happens as if there were only two virtual wheels situated at the middle of the axles.

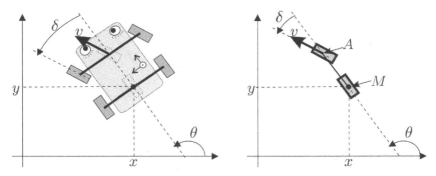

Figure 1.8. *Car moving on a plane (view from above)*

EXERCISE 1.12.– Car-trailer system

Let us consider the car studied in Exercise 1.11. Let us add a trailer to this car whose attachment point is found in the middle of the rear axle of the car, as illustrated by Figure 1.9. Find the state equations of the car-trailer system.

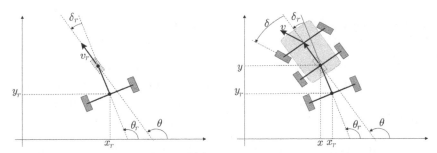

Figure 1.9. *Car with a trailer*

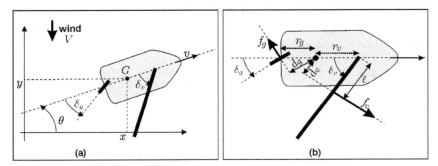

Figure 1.10. *Sailboat to be modeled*

EXERCISE 1.13.– Sailboat

Let us consider the sailboat to be modeled represented in Figure 1.10. The state vector x = $(x, y, \theta, \delta_v, \delta_g, v, \omega)^{\mathrm{T}}$, of dimension 7, is composed of the coordinates x, y of the center of gravity G of the boat (the drift is found at G), of the orientation θ, the angle δ_v of the sail, the angle δ_g of the rudder, the velocity v of the center of gravity G and of the angular velocity ω of the boat. The inputs u_1 and u_2 of the system are the derivatives of the angles δ_v and δ_g. The parameters (assumed to be known and constant) are: V the velocity of the wind, r_g the distance of the rudder to G, r_v the distance of the mast to G, α_g the lift of the rudder (if the rudder is perpendicular to the vessel's navigation, the water exerts a force of $\alpha_g v$ on the rudder), α_v the lift of the sail (if the sail is stationary and perpendicular to the wind, the latter exerts a force of $\alpha_v V$), α_f the coefficient of friction of the boat on the water in the direction of navigation (the water exerts a force opposite to the direction of navigation on the boat equal to $\alpha_f v^2$), α_θ the angular coefficient of friction (the water exerts a momentum of friction on the boat equal to $-\alpha_\theta \omega$); given the form of the boat that is rather streamlined in order to maintain its course, α_θ will be large compared to α_f), J the inertial momentum of the boat, ℓ the distance between the center of pressure of the sail and the mast, β the coefficient of drift (when the sail is released, the boat tends to drift in the direction of the wind with a velocity equal to βV). The state vector is composed of the coordinates of position, i.e. the coordinates x, y of the inertial center of the boat, the orientation θ, and the angles δ_v and δ_g of the sail and the rudder and the kinetic coordinates v and ω representing, respectively, the velocity of the center of rotation G and the angular velocity of the vessel. Find the state equations for our system $\dot{\mathbf{x}} = \mathbf{f}(\mathbf{x}, \mathbf{u})$, where $\mathbf{x} = (x, y, \theta, \delta_v, \delta_g, v, \omega)^T$ and $\mathbf{u} = (u_1, u_2)^T$.

EXERCISE 1.14.– Direct current motor

A direct current motor can be described by Figure 1.11, in which u is the supply voltage of the motor, i is the current absorbed by the motor, R is the armature resistance, L is the armature inductance, e is the electromotive force, ρ is the coefficient of friction in the motor, ω is the angular velocity of the motor and T_r is the torque exerted by the motor on the load.

Let us recall the equations of an ideal direct current motor: $e = K\Phi\omega$ and $T = K\Phi i$. In the case of an induction-independent motor, or a motor with permanent magnets, the flow Φ is constant. We are going to put ourselves in this situation.

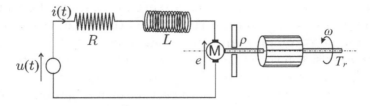

Figure 1.11. *Direct current motor*

1) We take as inputs of the system T_r and u. Find the state equations.

2) We connect a ventilator to the output of the system with a characteristic of $T_r = \alpha \omega^2$. Give the new state equations of the motor.

EXERCISE 1.15.– Electrical circuit

The electrical circuit of Figure 1.12 has the input as voltage $u(t)$ and the output as voltage $y(t)$. Find the state equations of the system. Is this a linear system?

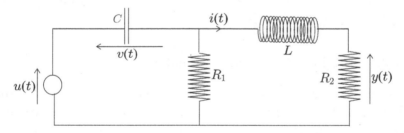

Figure 1.12. *Electrical circuit to be modeled*

EXERCISE 1.16.– The three containers

1) Let us consider two containers placed as shown in Figure 1.13.

In the left container, the water flows without friction in the direction of the right container. In the left container, the water flows in a fluid way, as opposed to the right container, where there are turbulences. These are turbulences that absorb the kinetic energy of the water and transform it into heat. Without these turbulences, we would have a perpetual back-and-forth movement of the water between the two containers. If a is the cross-section of the canal, then it shows the so-called *Torricelli*'s law that states that the water flow from the right container to the left one is equal to:

$$Q_D = a.sign\,(z_A - z_B)\,\sqrt{2g|z_A - z_B|}$$

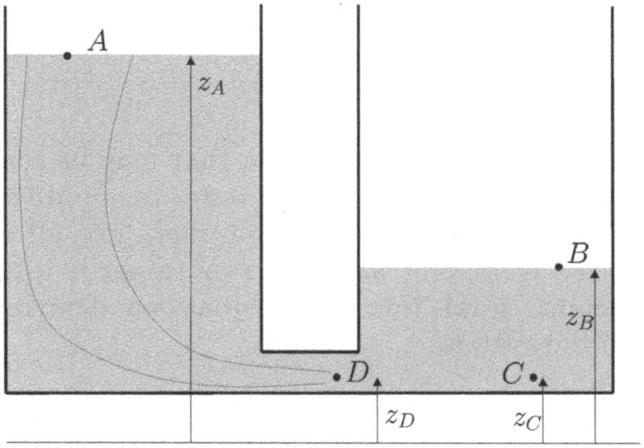

Figure 1.13. *Hydraulic system composed of two containers filled with water and connected with a canal*

2) Let us now consider the system composed of three containers as represented in Figure 1.14.

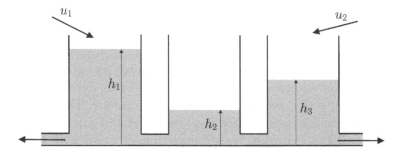

Figure 1.14. *System composed of three containers filled with water and connected with two canals*

The water from containers 1 and 3 can flow toward container 2, but also toward the outside with atmospheric pressure. The associated flow rates are given, following Toricelli's relation, by:

$$Q_{1\text{ext}} = a.\sqrt{2gh_1}$$
$$Q_{3\text{ext}} = a.\sqrt{2gh_3}$$

Similarly, the flow rate from a container i toward a container j is given by:

$$Q_{ij} = a.sign\,(h_i - h_j)\sqrt{2g|h_i - h_j|}$$

The state variables of this system that may be considered are the heights of the containers. In order to simplify, we will assume that the surfaces of the containers are all equal to 1 m^2; thus, the volume of water in a container is interlinked with its height. Find the state equations describing the dynamics of the system.

EXERCISE 1.17.– Pneumatic cylinder

Let us consider the pneumatic cylinder with return spring as shown in Figure 1.15. Such a cylinder is often referred to as single-acting since the air under pressure is only found in one of the two chambers.

The parameters of this system are the stiffness of the spring k, the surface of the piston a and the mass m at the end of the piston (the masses of all the other objects are ignored). We assume that everything happens under a constant temperature T_0. We will take as state vector $\mathbf{x} = (z, \dot{z}, p)$ where z is the position of the cylinder, $\dot{z}$ its velocity and p the pressure inside the chamber. The input of the system is the volumetric flow rate u of the air toward the chamber of the cylinder. In order to simplify, we will assume that there is vacuum in the spring chamber and that when

$z = 0$ (the cylinder is in the left hand limit) the spring is in equilibrium. Find the state equations of the pneumatic cylinder.

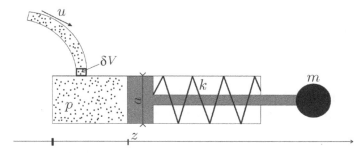

Figure 1.15. *Single-acting pneumatic cylinder*

EXERCISE 1.18.– Fibonacci sequence

We will now study the evolution of the number $y(k)$ of rabbit couples on a farm as a function of the year k. At year 0, there is only a single couple of newborn rabbits on the farm (and thus $y(0) = 1$). The rabbits only become fertile a year after their birth. It follows that at year 1, there is still a single couple of rabbits, but this couple is fertile (and thus $y(1) = 1$). A fertile couple gives birth, each year, to another couple of rabbits. Thus, at year 2, there is a fertile couple of rabbits and a newborn couple. This evolution can be described in Table 1.1, where N means *newborn* and A means *adult*.

$k=0$	$k=1$	$k=2$	$k=3$	$k=4$
N	A	A	A	A
		N	A	A
			N	A
				N
				N

Table 1.1. *Evolution of the number of rabbits*

Let us denote by $x_1(k)$ the number of newborn couples, by $x_2(k)$ the number of fertile couples and by $y(k)$ the total number of couples.

1) Give the state equations that govern the system.

2) Give the recurrence relation satisfied by $y(k)$.

EXERCISE 1.19.– Bus network

We consider a public transport system of buses with 4 lines and 4 buses. There are only two stations where travelers can change lines. This system can be represented by a Petri net (see Figure 1.16). Each token corresponds to a bus. The places p_1, p_2, p_3, p_4 represent the lines. These places are composed of a number that corresponds to the minimum amount of time that the token must remain in its place (this corresponds to the transit time). The transitions t_1, t_2 ensure synchronization. They are only crossed when each upstream place of the transition has at least one token that has waited sufficiently long. In this case, the upstream places lose a token and the downstream places gain one. This structure ensures that the correspondence will be performed systematically and that the buses will leave in pairs.

1) Let us assume that at time $t = 0$, the transitions t_1 and t_2 are crossed for the first time and that we are in the configuration of Figure 1.16 (this corresponds to the initialization). Give the crossing times for each of the transitions.

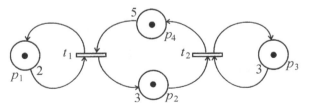

Figure 1.16. *Petri net of the bus system*

2) Let us denote by $x_i(k)$ the time when the transition t_i is crossed for the kth time. Show that the dynamics of the model can be written using states:

$$\mathbf{x}(k+1) = \mathbf{f}(\mathbf{x}(k))$$

where $\mathbf{x} = (x_1, x_2)^{\mathrm{T}}$ is the state vector. Remember that here k is not the time, but an event number.

3) Let us now attempt to reformulate elementary algebra by redefining the addition and multiplication operators (see [BAC 92]) as follows:

$$\begin{cases} a \oplus b = \max(a, b) \\ a \otimes b = a + b \end{cases}$$

Thus, $2 \oplus 3 = 3$, whereas $2 \otimes 3 = 5$. Show that in this new algebra (called max-plus), the previous system is linear.

4) Why do you think that matrix calculus (such as we know it) could not be used easily in this new algebra?

1.5. Solutions

Solution to Exercise 1.1 (integrator)

One possible state representation for the integrator is the following:

$$\begin{cases} \dot{x}(t) = u(t) \\ y(t) = x(t) \end{cases}$$

The matrices associated with this system are $\mathbf{A} = (0)$, $\mathbf{B} = (1)$, $\mathbf{C} = (1)$ and $\mathbf{D} = (0)$. They all have dimension 1×1.

Solution to Exercise 1.2 (second-order system)

By taking $\mathbf{x} = (y, \dot{y})$, this differential equation can be written in the form:

$$\begin{cases} \begin{pmatrix} \dot{x}_1 \\ \dot{x}_2 \end{pmatrix} = \begin{pmatrix} x_2 \\ -a_1 x_2 - a_0 x_1 + bu \end{pmatrix} \\ y = x_1 \end{cases}$$

or, in a standard form using state matrices **A**, **B**, **C** and **D**:

$$\begin{cases} \dot{\mathbf{x}} = \begin{pmatrix} 0 & 1 \\ -a_0 & -a_1 \end{pmatrix} \mathbf{x} + \begin{pmatrix} 0 \\ b \end{pmatrix} u \\ y = \begin{pmatrix} 1 & 0 \end{pmatrix} \mathbf{x} \end{cases}$$

Solution to Exercise 1.3 (mass-spring system)

1) The fundamental principle of dynamics applied to carriage 1 then to carriage 2 gives:

$$\begin{cases} -k_1 q_1 - \alpha \dot{q}_1 + k_2(q_2 - q_1) = m_1 \ddot{q}_1 \\ u - \alpha \dot{q}_2 - k_2(q_2 - q_1) = m_2 \ddot{q}_2 \end{cases}$$

In other words:

$$\begin{pmatrix} \ddot{q}_1 \\ \ddot{q}_2 \end{pmatrix} = \begin{pmatrix} \frac{1}{m_1}\left(-(k_1 + k_2)q_1 + k_2 q_2 - \alpha \dot{q}_1\right) \\ \frac{1}{m_2}(k_2 q_1 - k_2 q_2 - \alpha \dot{q}_2 + u) \end{pmatrix}$$

We thus have the following state representation:

$$\begin{cases} \dot{\mathbf{x}} = \begin{pmatrix} 0 & 0 & 1 & 0 \\ 0 & 0 & 0 & 1 \\ -\frac{k_1+k_2}{m_1} & \frac{k_2}{m_1} & -\frac{\alpha}{m_1} & 0 \\ \frac{k_2}{m_2} & -\frac{k_2}{m_2} & 0 & -\frac{\alpha}{m_2} \end{pmatrix} \mathbf{x} + \begin{pmatrix} 0 \\ 0 \\ 0 \\ \frac{1}{m_2} \end{pmatrix} u \\ q_1 = \begin{pmatrix} 1 & 0 & 0 & 0 \end{pmatrix} \mathbf{x} \end{cases}$$

2) Yes, this system is linear since it can be written as:

$$\begin{cases} \dot{\mathbf{x}} = \mathbf{A}\mathbf{x} + \mathbf{B}\mathbf{u} \\ \mathbf{y} = \mathbf{C}\mathbf{x} + \mathbf{D}\mathbf{u} \end{cases}$$

Solution to Exercise 1.4 (simple pendulum)

1) Following the fundamental principle of dynamics, we have:

$$-\ell m g \sin q + u = J \ddot{q}$$

where ℓ is the length of the pendulum. However, for our example, $J = m\ell^2$, therefore:

$$\ddot{q} = \frac{u - \ell m g \sin q}{m\ell^2}$$

Let us take as state vector $\mathbf{x} = (q, \dot{q})$. The state equations of the system are then written as:

$$\begin{aligned} \frac{d}{dt}\begin{pmatrix} q \\ \dot{q} \end{pmatrix} &= \begin{pmatrix} \dot{q} \\ \frac{u - \ell m g \sin q}{m\ell^2} \end{pmatrix} \\ y &= \ell \sin q \end{aligned}$$

or:

$$\begin{aligned} \begin{pmatrix} \dot{x}_1 \\ \dot{x}_2 \end{pmatrix} &= \begin{pmatrix} x_2 \\ \frac{u - \ell m g \sin x_1}{m\ell^2} \end{pmatrix} \\ y &= \ell \sin x_1 \end{aligned}$$

2) The mechanical energy of the pendulum is given by:

$$E_m = \underbrace{\frac{1}{2} m \ell^2 \dot{q}^2}_{\text{kinetic energy}} + \underbrace{mg\ell (1 - \cos q)}_{\text{potential energy}}$$

When the torque u is nil, we have:

$$\frac{dE_m}{dt} = \tfrac{1}{2}m\ell^2\left(2\dot{q}\ddot{q}\right) + mg\ell\dot{q}\sin q$$
$$= m\ell^2\left(\dot{q}\frac{-\ell mg\sin q}{m\ell^2}\right) + mg\ell\dot{q}\sin q = 0$$

The mechanical energy of the pendulum therefore remains constant, which is coherent with the fact that the pendulum without friction is a conservative system.

Solution to Exercise 1.5 (dynamic modeling of an inverted rod pendulum)

1) The fundamental principle of dynamics applied to the carriage and the pendulum gives us:

$$\begin{aligned}
(u - R_x)\vec{i} &= M\ddot{x}\vec{i} \quad \text{(carriage in translation)} \\
R_x\vec{i} + R_y\vec{j} - mg\vec{j} &= m\dot{\mathbf{v}}_B \quad \text{(pendulum in translation)} \\
R_x\ell\cos\theta + R_y\ell\sin\theta &= 0\ddot{\theta} \quad \text{(pendulum in rotation)}
\end{aligned}$$

where $\mathbf{v}_B$ is the velocity vector of point B. For the third equation, the inertial momentum of the pendulum was defined nil.

2) Since:

$$\overrightarrow{OB} = (x - \ell\sin\theta)\vec{i} + \ell\cos\theta\vec{j}$$

we have:

$$\mathbf{v}_B = \left(\dot{x} - \ell\dot{\theta}\cos\theta\right)\vec{i} - \ell\dot{\theta}\sin\theta\vec{j}$$

Therefore, the acceleration of point B is given by:

$$\dot{\mathbf{v}}_B = \left(\ddot{x} - \ell\ddot{\theta}\cos\theta + \ell\dot{\theta}^2\sin\theta\right)\vec{i} - \left(\ell\ddot{\theta}\sin\theta + \ell\dot{\theta}^2\cos\theta\right)\vec{j}$$

3) It is the vector of the degrees of freedom and of their derivatives. There are no non-holonomic constraints.

4) After scalar decomposition of the dynamics equations given above, we obtain:

$$\begin{cases} M\ddot{x} = u - R_x & \text{(i)} \\ R_x = m\left(\ddot{x} - \ell\ddot{\theta}\cos\theta + \ell\dot{\theta}^2\sin\theta\right) & \text{(ii)} \\ R_y - mg = -m\left(\ell\ddot{\theta}\sin\theta + \ell\dot{\theta}^2\cos\theta\right) & \text{(iii)} \\ R_x\cos\theta + R_y\sin\theta = 0 & \text{(iv)} \end{cases}$$

These four equations describe, respectively, (i) the carriage in translation, (ii) the pendulum in translation following $\vec{i}$, (iii) the pendulum in translation following $\vec{j}$ and (iv) the pendulum in rotation. We thus verify that the number of degrees of freedom (here x and θ) added to the number of internal forces (here R_x and R_y) is equal to the number of equations. In a matrix form, these equations become:

$$\begin{pmatrix} M & 0 & 1 & 0 \\ -m & m\ell\cos\theta & 1 & 0 \\ 0 & m\ell\sin\theta & 0 & 1 \\ 0 & 0 & \cos\theta & \sin\theta \end{pmatrix} \begin{pmatrix} \ddot{x} \\ \ddot{\theta} \\ R_x \\ R_y \end{pmatrix} = \begin{pmatrix} u \\ m\ell\dot{\theta}^2\sin\theta \\ mg - m\ell\dot{\theta}^2\cos\theta \\ 0 \end{pmatrix}$$

Therefore:

$$\begin{pmatrix} \ddot{x} \\ \ddot{\theta} \end{pmatrix} = \begin{pmatrix} 1 & 0 & 0 & 0 \\ 0 & 1 & 0 & 0 \end{pmatrix} \cdot \begin{pmatrix} M & 0 & 1 & 0 \\ -m & m\ell\cos\theta & 1 & 0 \\ 0 & m\ell\sin\theta & 0 & 1 \\ 0 & 0 & \cos\theta & \sin\theta \end{pmatrix}^{-1} \cdot \begin{pmatrix} u \\ m\ell\dot{\theta}^2\sin\theta \\ mg - m\ell\dot{\theta}^2\cos\theta \\ 0 \end{pmatrix}$$

$$= \begin{pmatrix} \frac{-m\sin\theta(\ell\dot{\theta}^2 - g\cos\theta) + u}{M + m\sin^2\theta} \\ \frac{\sin\theta((M+m)g - m\ell\dot{\theta}^2\cos\theta) + \cos\theta\, u}{\ell(M + m\sin^2\theta)} \end{pmatrix}$$

The state equations are therefore written as:

$$\frac{d}{dt}\begin{pmatrix} x \\ \theta \\ \dot{x} \\ \dot{\theta} \end{pmatrix} = \begin{pmatrix} \dot{x} \\ \dot{\theta} \\ \frac{-m\sin\theta(\ell\dot{\theta}^2 - g\cos\theta)}{M + m\sin^2\theta} \\ \frac{(\sin\theta)((M+m)g - m\ell\dot{\theta}^2\cos\theta)}{\ell(M + m\sin^2\theta)} \end{pmatrix} + \begin{pmatrix} 0 \\ 0 \\ \frac{1}{M + m\sin^2\theta} \\ \frac{\cos\theta}{\ell(M + m\sin^2\theta)} \end{pmatrix} u$$

or equivalently:

$$\frac{d}{dt}\begin{pmatrix} x_1 \\ x_2 \\ x_3 \\ x_4 \end{pmatrix} = \begin{pmatrix} x_3 \\ x_4 \\ \frac{-m \sin x_2 (\ell x_4^2 - g \cos x_2)}{M + m \sin^2 x_2} \\ \frac{(\sin x_2)((M+m)g - m\ell x_4^2 \cos x_2)}{\ell(M + m \sin^2 x_2)} \end{pmatrix} + \begin{pmatrix} 0 \\ 0 \\ \frac{1}{M + m \sin^2 x_2} \\ \frac{\cos x_2}{\ell(M + m \sin^2 x_2)} \end{pmatrix} u$$

Solution to Exercise 1.6 (kinematic modeling of an inverted rod pendulum)

1) Following the dynamic model of the inverted rod pendulum, we obtain:

$$a = \frac{1}{M + m \sin^2 \theta} \left(-m \sin \theta (\ell \dot{\theta}^2 - g \cos \theta) + u \right)$$

After isolating u, we get:

$$u = m \sin \theta \left(\ell \dot{\theta}^2 - g \cos \theta \right) + \left(M + m \sin^2 \theta \right) a$$

Therefore:

$$\begin{aligned} \ddot{\theta} &= \frac{\sin \theta ((M+m)g - m\ell \dot{\theta}^2 \cos \theta)}{\ell(M + m \sin^2 \theta)} + \frac{\cos \theta}{\ell(M + m \sin^2 \theta)} u \\ &= \frac{\sin \theta ((M+m)g - m\ell \dot{\theta}^2 \cos \theta)}{\ell(M + m \sin^2 \theta)} + \frac{\cos \theta}{\ell(M + m \sin^2 \theta)} \left(m \sin \theta \left(\ell \dot{\theta}^2 - g \cos \theta \right) + (M + m \sin^2 \theta) a \right) \\ &= \frac{1}{\ell(M + m \sin^2 \theta)} \left((M+m)g \sin \theta - gm \sin \theta \cos^2 \theta + (M + m \sin^2 \theta) \cos \theta a \right) \\ &= \frac{g \sin \theta}{\ell} + \frac{\cos \theta}{\ell} a \end{aligned}$$

Let us note that this relation could have been obtained directly by noticing that:

$$\ell \ddot{\theta} = \underbrace{a . \cos \theta}_{\text{acceleration of } A \text{ that contributes to the rotation}}$$
$$+ \underbrace{g . \sin \theta}_{\text{acceleration of } B \text{ from the viewpoint of } A}$$

REMARK.– In order to obtain this relation in a more rigorous manner and without the use of the dynamic model, we would need to write the temporal derivative of the formula of the composition of the velocities (or *Varignon*'s formula), in other words:

$$\dot{\vec{v}}_A = \dot{\vec{v}}_B + \overrightarrow{AB} \wedge \vec{\omega}$$

and write this formula in the frame of the pendulum. We obtain:

$$\begin{pmatrix} a\cos\theta \\ -a\sin\theta \\ 0 \end{pmatrix} = \begin{pmatrix} -g\sin\theta \\ n \\ 0 \end{pmatrix} + \begin{pmatrix} 0 \\ \ell \\ 0 \end{pmatrix} \wedge \begin{pmatrix} 0 \\ 0 \\ \dot{\omega} \end{pmatrix}$$

where n corresponds to the normal acceleration of the mass m. We thus obtain, in addition to the desired relation, the normal acceleration $n = -a\sin\theta$ that will not be used.

The model therefore becomes:

$$\frac{d}{dt}\begin{pmatrix} x \\ \theta \\ \dot{x} \\ \dot{\theta} \end{pmatrix} = \begin{pmatrix} \dot{x} \\ \dot{\theta} \\ 0 \\ \frac{g\sin\theta}{\ell} \end{pmatrix} + \begin{pmatrix} 0 \\ 0 \\ 1 \\ \frac{\cos\theta}{\ell} \end{pmatrix} a$$

This model, referred to as *kinetic*, only involves positions, velocities and accelerations. It is much more simple than the dynamic model and contains less coefficients. On the other hand, it corresponds less to reality since the real input is a force and not an acceleration.

2) In the case of the inverted rod pendulum, we can move from the dynamic model with input u to a kinetic model with input a by generating u with a *proportional control* of type *high gain* of the form:

$$u = K(a - \ddot{x})$$

with K very large and where a is a new input (see Figure 1.17).

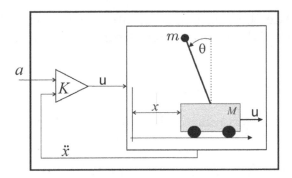

Figure 1.17. *The inverted rod pendulum, looped by a high gain K, behaves like a kinematic model*

The acceleration $\ddot{x}$ can be measured by an accelerometer. If K is sufficiently large, we will have a control u that will create the desired acceleration a, in other words we will be able to say that $\ddot{x} = a$. Thus, the system could be described by state equations of the kinematic model that do not involve any of the inertial parameters of the system. A controller that will be designed over the kinematic model will therefore be more robust than that designed over the dynamic model since the controller will work no matter what the masses (m, M), the inertial moments, the frictions, etc. are. This high gain-type control is very close to the op-amp principle. In addition to being more robust, such an approach also allows to have a simpler model that is easier to obtain.

Solution to Exercise 1.7 (segway)

In order to find the state equations, we apply the fundamental principle of dynamics on each subsystem, more precisely the wheel and the body. We have:

$$\begin{cases} -R_x + F_x = -Ma\ddot{\alpha} & \text{(wheel in translation)} \\ F_x a + u = J_M \ddot{\alpha} & \text{(wheel in rotation)} \\ R_x \vec{i} + R_y \vec{j} - mg\vec{j} = m\dot{\vec{v}}_B & \text{(body in translation)} \\ R_x \ell \cos\theta + R_y \ell \sin\theta - u = J_p \ddot{\theta} & \text{(body in rotation)} \end{cases}$$

where $\mathbf{v}_B$ is the velocity vector of point B. Since:

$$\overrightarrow{OB} = (-a\alpha - \ell \sin\theta)\vec{i} + (\ell \cos\theta + a)\vec{j}$$

using differentiation, we obtain:

$$\mathbf{v}_B = \left(-a\dot\alpha - \ell\dot\theta \cos\theta\right)\vec{i} - \ell\dot\theta \sin\theta \vec{j}$$

or:

$$\dot{\mathbf{v}}_B = \left(-a\ddot\alpha - \ell\ddot\theta \cos\theta + \ell\dot\theta^2 \sin\theta\right)\vec{i} - \left(\ell\ddot\theta \sin\theta + \ell\dot\theta^2 \cos\theta\right)\vec{j}$$

Thus, after scalar decomposition, the dynamics equations become:

$$\begin{cases} -R_x + F_x = -Ma\ddot\alpha \\ F_x a + u = J_M \ddot\alpha \\ R_x = m(-a\ddot\alpha - \ell\ddot\theta \cos\theta + \ell\dot\theta^2 \sin\theta) \\ R_y - mg = -m(\ell\ddot\theta \sin\theta + \ell\dot\theta^2 \cos\theta) \\ R_x \ell \cos\theta + R_y \ell \sin\theta - u = J_p \ddot\theta \end{cases}$$

We verify that the number of degrees of freedom (here α and θ) added to the number of components of the internal forces (here R_x, R_y and F_x) is equal to the number of equations. In matrix form, these equations become:

$$\begin{pmatrix} Ma & 0 & -1 & 0 & 1 \\ J_M & 0 & 0 & 0 & -a \\ ma & m\ell \cos\theta & 1 & 0 & 0 \\ 0 & m\ell \sin\theta & 0 & 1 & 0 \\ 0 & -J_p & \ell \cos\theta & \ell \sin\theta & 0 \end{pmatrix} \begin{pmatrix} \ddot\alpha \\ \ddot\theta \\ R_x \\ R_y \\ F_x \end{pmatrix} = \begin{pmatrix} 0 \\ u \\ m\ell\dot\theta^2 \sin\theta \\ mg - m\ell\dot\theta^2 \cos\theta \\ u \end{pmatrix}$$

Therefore:

$$\begin{pmatrix} \ddot\alpha \\ \ddot\theta \end{pmatrix} = \begin{pmatrix} 1 & 0 & 0 & 0 & 0 \\ 0 & 1 & 0 & 0 & 0 \end{pmatrix} \cdot \begin{pmatrix} Ma & 0 & -1 & 0 & 1 \\ J_M & 0 & 0 & 0 & -a \\ ma & m\ell \cos\theta & 1 & 0 & 0 \\ 0 & m\ell \sin\theta & 0 & 1 & 0 \\ 0 & -J_p & \ell \cos\theta & \ell \sin\theta & 0 \end{pmatrix}^{-1} \cdot \begin{pmatrix} 0 \\ u \\ m\ell\dot\theta^2 \sin\theta \\ mg - m\ell\dot\theta^2 \cos\theta \\ u \end{pmatrix}$$

In other words:
$$\begin{cases} \ddot{\alpha} = \frac{\mu_3\left(\mu_2\dot{\theta}^2 - \mu_g\cos\theta\right)\sin\theta + (\mu_2+\mu_3\cos\theta)u}{\mu_1\mu_2 - \mu_3^2\cos^2\theta} \\ \ddot{\theta} = \frac{\left(\mu_1\mu_g - \mu_3^2\dot{\theta}^2\cos\theta\right)\sin\theta - (\mu_1+\mu_3\cos\theta)u}{\mu_1\mu_2 - \mu_3^2\cos^2\theta} \end{cases}$$

with:
$$\mu_1 = J_M + a^2(m+M), \quad \mu_2 = J_p + m\ell^2,$$
$$\mu_3 = am\ell, \qquad \mu_g = g\ell m$$

The state equations are therefore written as:

$$\frac{d}{dt}\begin{pmatrix}\alpha \\ \theta \\ \dot{\alpha} \\ \dot{\theta}\end{pmatrix} = \begin{pmatrix} \dot{\alpha} \\ \dot{\theta} \\ \frac{\mu_3\left(\mu_2\dot{\theta}^2 - \mu_g\cos\theta\right)\sin\theta + (\mu_2+\mu_3\cos\theta)u}{\mu_1\mu_2 - \mu_3^2\cos^2\theta} \\ \frac{\left(\mu_1\mu_g - \mu_3^2\dot{\theta}^2\cos\theta\right)\sin\theta - (\mu_1+\mu_3\cos\theta)u}{\mu_1\mu_2 - \mu_3^2\cos^2\theta} \end{pmatrix}$$

By taking $\mathbf{x} = \left(\alpha, \theta, \dot{\alpha}, \dot{\theta}\right)^{\mathrm{T}}$, these equations become:

$$\begin{pmatrix}\dot{x}_1 \\ \dot{x}_2 \\ \dot{x}_3 \\ \dot{x}_4\end{pmatrix} = \begin{pmatrix} x_3 \\ x_4 \\ \frac{\mu_3\left(\mu_2^2 x_4 - \mu_g\cos x_2\right)\sin x_2 + (\mu_2+\mu_3\cos x_2)u}{\mu_1\mu_2 - \mu_3^2\cos^2 x_2} \\ \frac{\left(\mu_1\mu_g - \mu_3^2 x_4^2\cos x_2\right)\sin x_2 - (\mu_1+\mu_3\cos x_2)u}{\mu_1\mu_2 - \mu_3^2\cos^2 x_2} \end{pmatrix}$$

Solution to Exercise 1.8 (Hamilton's method)

1) The Hamiltonian is written as:

$$H(q,p) = \underbrace{\frac{1}{2}m(\ell\dot{q})^2}_{\text{Kinetic energy}} + \underbrace{mg\ell(1-\cos q)}_{\text{Potential energy}}$$

$$= \frac{1}{2}\frac{p^2}{m\ell^2} + mg\ell(1-\cos q)$$

since the amount of movement of the pendulum (or rather the kinetic moment in this case) is $p = J\dot{q} = m\ell^2\dot{q}$ and thus $m(\ell\dot{q})^2 = m\left(\ell\frac{p}{m\ell^2}\right)^2 = \frac{p^2}{m\ell^2}$. The state equations of the pendulum are therefore:

$$\begin{cases} \dot{q} = \frac{\partial H(q,p)}{\partial p} = \frac{p}{m\ell^2} \\ \dot{p} = -\frac{\partial H(q,p)}{\partial q} = -mg\ell\sin q \end{cases}$$

where the state vector here is $x = (q,p)^T$. Let us note that:

$$\ddot{q} = \frac{\dot{p}}{m\ell^2} = \frac{-mg\ell\sin q}{m\ell^2} = \frac{-g\sin q}{\ell}$$

and we come back to the differential equation of the pendulum.

2) We have:

$$\dot{H} = \frac{\partial H(\mathbf{q},\mathbf{p})}{\partial \mathbf{p}}\dot{\mathbf{p}} + \frac{\partial H(\mathbf{q},\mathbf{p})}{\partial \mathbf{q}}\dot{\mathbf{q}} = 0$$

The Hamiltonian (or, equivalently, the mechanical energy) is thus constant.

Solution to Exercise 1.9 (omnidirectional robot)

1) We have $r\omega_i = \langle \mathbf{v}_i, \mathbf{i}_i \rangle$. However, following the velocity composition formula (Varignon's formula), $\mathbf{v}_i = \mathbf{v} - a\dot{\theta}\mathbf{i}_{i_\perp}$. Therefore:

$$r\omega_i = \langle \mathbf{v} - a\dot{\theta}.\mathbf{i}_{i_\perp}, \mathbf{i}_i \rangle = \langle \mathbf{v}, \mathbf{i}_i \rangle - a\dot{\theta}$$

or:

$$\mathbf{v} = \begin{pmatrix} \dot{x} \\ \dot{y} \end{pmatrix}, \mathbf{i}_1 = \begin{pmatrix} -\sin\theta \\ \cos\theta \end{pmatrix}, \mathbf{i}_2 = \begin{pmatrix} -\sin\left(\theta - \frac{\pi}{3}\right) \\ \cos\left(\theta - \frac{\pi}{3}\right) \end{pmatrix},$$

$$\mathbf{i}_3 = \begin{pmatrix} -\sin\left(\theta + \frac{\pi}{3}\right) \\ \cos\left(\theta + \frac{\pi}{3}\right) \end{pmatrix}$$

Therefore:

$$\begin{pmatrix} \omega_1 \\ \omega_2 \\ \omega_3 \end{pmatrix} = \frac{1}{r} \underbrace{\begin{pmatrix} -\sin\theta & \cos\theta & -a \\ -\sin\left(\theta - \frac{\pi}{3}\right) & \cos\left(\theta - \frac{\pi}{3}\right) & -a \\ -\sin\left(\theta + \frac{\pi}{3}\right) & \cos\left(\theta + \frac{\pi}{3}\right) & -a \end{pmatrix}}_{\mathbf{A}(\theta)} \begin{pmatrix} \dot{x} \\ \dot{y} \\ \dot{\theta} \end{pmatrix}$$

The state equations are therefore:

$$\begin{pmatrix} \dot{x} \\ \dot{y} \\ \dot{\theta} \end{pmatrix} = \mathbf{A}^{-1}(\theta).\omega$$

2) For this we need to design an input controller $\mathbf{u} = (u_1, u_2)^{\mathrm{T}}$ and an output controller $\omega = (\omega_1, \omega_2, \omega_3)^{\mathrm{T}}$. The new inputs u_1, u_2 correspond to the desired angular velocity and acceleration. Let us choose the controller:

$$\begin{cases} \dot{v} = u_2 \\ \omega = \mathbf{A}(\theta) . \begin{pmatrix} v\cos\theta \\ v\sin\theta \\ u_1 \end{pmatrix} \end{cases}$$

The loop is represented in Figure 1.18.

The loop system is then written as:

$$\begin{cases} \dot{x} = v\cos\theta \\ \dot{y} = v\sin\theta \\ \dot{\theta} = u_1 \\ \dot{v} = u_2 \end{cases}$$

Solution to Exercise 1.10 (modeling a tank)

1) The state vector cannot be chosen to be equal to $(x, y, \theta, \dot{x}, \dot{y}, \dot{\theta})^{\mathrm{T}}$, which would seem natural with respect to

Lagrangian theory. Indeed, if this was our choice, some states would have no physical meaning. For instance the state:

$$\left(x = 0, y = 0, \theta = 0, \dot{x} = 1, \dot{y} = 1, \dot{\theta} = 0\right)$$

has no meaning since the tank is not allowed to skid. This phenomenon is due to the existence of wheels that creates constraints between the natural state variables. Here, we necessarily have the so-called *non-holonomic* constraint:

$$\dot{y} = \dot{x}\tan\theta$$

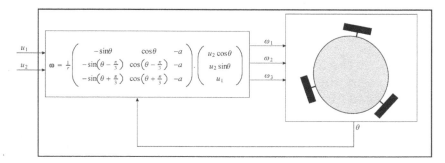

Figure 1.18. *The robot with omni wheels looped in this manner behaves like a tank*

Mechanical systems for which there are such constraints on the equality of natural state variables (by natural state variables we mean the vector $(\mathbf{q}, \dot{\mathbf{q}})$ where $\mathbf{q}$ is the vector of the degrees of liberty of our system) are said to be *non-holonomic*. When such a situation arises, it is useful to use these constraints in order to reduce the number of state variables and this, until no more constraints are left between the state variables.

2) This choice of state variables is easily understood in the sense that these variables allow us to draw the tank (x, y, θ) and the knowledge of v_1, v_2 allows us to calculate the variables $\dot{x}$, $\dot{y}$, $\dot{\theta}$. Moreover, every arbitrary choice of vector

(x, y, θ, v_1, v_2) corresponds to a physically possible situation. The state equations of the system are:

$$\begin{pmatrix} \dot{x} \\ \dot{y} \\ \dot{\theta} \\ \dot{v}_1 \\ \dot{v}_2 \end{pmatrix} = \begin{pmatrix} \frac{v_1+v_2}{2} \cos \theta \\ \frac{v_1+v_2}{2} \sin \theta \\ \frac{v_2-v_1}{\ell} \\ Ru_1 \\ Ru_2 \end{pmatrix}$$

where ℓ is the distance between the two wheels. The third relation on $\dot{\theta}$ is obtained by the velocity composition rule (Varignon's formula). Indeed, we have:

$$\mathbf{v}_2 = \mathbf{v}_1 + \overrightarrow{C_2 C_1} \wedge \vec{\omega}$$

where $\vec{\omega}$ is the instantaneous rotation vector of the tank and $\mathbf{v}_1$ and $\mathbf{v}_2$ are the velocity vectors of the centers of the wheels. Let us note that this relation is a vectorial relation that depends on the observer but that is independent of the frame. Let us express this in the frame of the tank, represented on the figure. We must be careful not to confuse the observer fixed on the ground with the frame in which the relation is expressed. This equation is written as:

$$\underbrace{\begin{pmatrix} v_2 \\ 0 \\ 0 \end{pmatrix}}_{\mathbf{v}_2} = \underbrace{\begin{pmatrix} v_1 \\ 0 \\ 0 \end{pmatrix}}_{\mathbf{v}_1} + \underbrace{\begin{pmatrix} 0 \\ \ell \\ 0 \end{pmatrix}}_{\overrightarrow{C_2 C_1}} \wedge \underbrace{\begin{pmatrix} 0 \\ 0 \\ \dot{\theta} \end{pmatrix}}_{\vec{\omega}}$$

We thus obtain $v_2 = v_1 + \ell \dot{\theta}$ or $\dot{\theta} = \frac{v_2-v_1}{\ell}$.

Solution to Exercise 1.11 (modeling a car)

Let us consider an observer fixed with respect to the ground. Following the velocity composition rule (Varignon's formula), we have:

$$\mathbf{v}_A = \mathbf{v}_M + \overrightarrow{AM} \wedge \vec{\omega}$$

where $\vec{\omega}$ is the instantaneous rotation vector of the car. Let us express this vectorial relation in the frame of the car, which is represented in the figure:

$$\begin{pmatrix} v\cos\delta \\ v\sin\delta \\ 0 \end{pmatrix} = \begin{pmatrix} v_M \\ 0 \\ 0 \end{pmatrix} + \begin{pmatrix} -L \\ 0 \\ 0 \end{pmatrix} \wedge \begin{pmatrix} 0 \\ 0 \\ \dot{\theta} \end{pmatrix}$$

where L is the distance between the front and rear axles. Therefore:

$$\begin{pmatrix} v\cos\delta \\ v\sin\delta \end{pmatrix} = \begin{pmatrix} v_M \\ 0 \end{pmatrix} + \begin{pmatrix} 0 \\ L\dot{\theta} \end{pmatrix}$$

Thus:

$$\dot{\theta} = \frac{v\sin\delta}{L}$$

and:

$$\begin{cases} \dot{x} = v_M \cos\theta = v\cos\delta\cos\theta \\ \dot{y} = v_M \sin\theta = v\cos\delta\sin\theta \end{cases}$$

The evolution equation of the car is therefore written as:

$$\begin{pmatrix} \dot{x} \\ \dot{y} \\ \dot{\theta} \\ \dot{v} \\ \dot{\delta} \end{pmatrix} = \begin{pmatrix} v\cos\delta\cos\theta \\ v\cos\delta\sin\theta \\ \frac{v\sin\delta}{L} \\ u_1 \\ u_2 \end{pmatrix}$$

Solution to Exercise 1.12 (car-trailer system)

By looking at the figure and using the state equations of the car, we have:

$$\dot{\theta}_r = \frac{v_r \sin\delta_r}{L_r}$$

with:

$$v_r = \sqrt{\dot{x}^2 + \dot{y}^2} = \sqrt{(v\cos\delta\cos\theta)^2 + (v\cos\delta\sin\theta)^2} = v\cos\delta$$

$$\delta_r = \theta - \theta_r$$

The parameter L_r represents the distance between the attachment point and the middle of the axle of the trailer. However, only θ_r has to be added as state variable to those of the car. Indeed, it is clear that $x_r, y_r, v_r, \delta_r, \dot{x}_r, \dot{y}_r \ldots$ can be obtained analytically from the sole knowledge of the state of the car and the angle θ_r. Thus, the state equations of the cat-trailer system are given by:

$$\begin{pmatrix} \dot{x} \\ \dot{y} \\ \dot{\theta} \\ \dot{\theta}_r \\ \dot{v} \\ \dot{\delta} \end{pmatrix} = \begin{pmatrix} v\cos\delta\cos\theta \\ v\cos\delta\sin\theta \\ \frac{v\sin\delta}{L} \\ \frac{v\cos\delta\sin(\theta-\theta_r)}{L_r} \\ u_1 \\ u_2 \end{pmatrix}$$

Solution to Exercise 1.13 (sailboat)

The modeling that we will propose here is inspired from the article [JAU 04]. Even though it is simplistic, the obtained model remains relatively consistent with reality and is used for the simulation of robotic sailboats (such as Vaimos) in order to test the behavior of the controllers. In order to perform this modeling, we will use the fundamental principle of dynamics in translation (in order to obtain an expression of the tangential acceleration $\dot{v}$) and then in rotation (in order to obtain an expression of the angular acceleration $\dot{\omega}$).

TANGENTIAL ACCELERATION $\dot{v}$.– The wind exerts an orthogonal force on the sail with an intensity equal to:

$$f_v = \alpha_v \left(V\cos(\theta + \delta_v) - v\sin\delta_v \right)$$

Concerning the water, it exerts a force on the rudder that is equal to:

$$f_g = \alpha_g v \sin \delta_g$$

orthogonal to the rudder. The friction force exerted by it on the boat is assumed to be proportional to the square of the boat's velocity. The fundamental equation of dynamics, projected following the axis of the boat gives:

$$m\dot{v} = \sin \delta_v f_v - \sin \delta_g f_g - \alpha_f v^2$$

The radial acceleration can be considered as nil if we assume that the drift is perfect.

ANGULAR ACCELERATION $\dot{\omega}$.– Among the forces that act on the rotation of the boat, we can find the forces f_v and f_g exerted by the sail and the rudder, but also a force of angular friction which we assume to be viscous. The fundamental equation of dynamics gives us:

$$J\dot{\omega} = d_v f_v - d_g f_g - \alpha_\theta \omega$$

where:

$$\begin{cases} d_v = \ell - r_v \cos \delta_v \\ d_g = r_g \cos \delta_g \end{cases}$$

The state equations of the boat are therefore written as:

$$\begin{cases} \dot{x} = v \cos \theta & \text{(i)} \\ \dot{y} = v \sin \theta - \beta V & \text{(ii)} \\ \dot{\theta} = \omega & \text{(iii)} \\ \dot{\delta}_v = u_1 & \text{(iv)} \\ \dot{\delta}_g = u_2 & \text{(v)} \\ \dot{v} = \frac{f_v \sin \delta_v - f_g \sin \delta_g - \alpha_f v^2}{m} & \text{(vi)} \\ \dot{\omega} = \frac{(\ell - r_v \cos \delta_v) f_v - r_g \cos \delta_g f_g - \alpha_\theta \omega}{J} & \text{(vii)} \\ f_v = \alpha_v (V \cos(\theta + \delta_v) - v \sin \delta_v) & \text{(viii)} \\ f_g = \alpha_g v \sin \delta_g & \text{(ix)} \end{cases}$$

Let us note that these previous equations are not differential but algebraic ones. In order to be perfectly consistent with a state equation, we would need to remove these two equations as well as the two internal forces f_v and f_g that appear in equations (vi) and (vii).

Solution to Exercise 1.14 (direct current motor)

1) The equations governing the system are given by:

$u = Ri + L\frac{di}{dt} + e$ (electrical part)
$J\dot{\omega} = T - \rho\omega - T_r$ (mechanical part)

However, the equations of a direct current machine are $e = K\Phi\omega$ and $T = K\Phi i$. Therefore:

$u = Ri + L\frac{di}{dt} + K\Phi\omega$
$J\dot{\omega} = K\Phi i - \rho\omega - T_r$

We then have the following linear state equations:

$$\begin{cases} \frac{di}{dt} = -\frac{R}{L}i - \frac{K\Phi}{L}\omega + \frac{u}{L} \\ \dot{\omega} = \frac{K\Phi}{J}i - \frac{\rho}{J}\omega - \frac{T_r}{J} \end{cases}$$

where the inputs are u and T_r and the state variables are i and ω. These equations can be written in matrix form:

$$\frac{d}{dt}\begin{pmatrix} i \\ \omega \end{pmatrix} = \begin{pmatrix} -\frac{R}{L} & -\frac{K\Phi}{L} \\ \frac{K\Phi}{J} & -\frac{\rho}{J} \end{pmatrix}\begin{pmatrix} i \\ \omega \end{pmatrix} + \begin{pmatrix} \frac{1}{L} & 0 \\ 0 & -\frac{1}{J} \end{pmatrix}\begin{pmatrix} u \\ T_r \end{pmatrix}$$

2) When the motor is running, the torque T_r is being imposed. The following table gives some mechanical characteristics, in continuous output, of the pair (T_r, ω):

$T_r = C^{te}$ motor used for lifting
$T_r = \alpha\omega$ locomotive, mixer, pump
$T_r = \frac{\alpha}{\omega}$ machine tool (lathe, milling machine)
$T_r = \alpha\omega^2$ ventilator, fast car

In our case, $T_r = \alpha\omega^2$. The motor now only has a single input that is the armature voltage $u(t)$. We have:

$$\begin{cases} \frac{di}{dt} = -\frac{R}{L}i - \frac{K\Phi}{L}\omega + \frac{u}{L} \\ \dot{\omega} = \frac{K\Phi}{J}i - \frac{\rho}{J}\omega - \frac{\alpha\omega^2}{J} \end{cases}$$

This is a nonlinear system. It now only has a single input $u(t)$.

Solution to Exercise 1.15 (RLC circuit)

Let i_1 be the electrical current in the resistance R_1 (top to bottom). Following the node and mesh rules, we have:

$$\begin{cases} u(t) - v(t) - R_1 i_1(t) = 0 \text{ (mesh rule)} \\ L\frac{di}{dt} + R_2 i(t) - R_1 i_1(t) = 0 \text{ (mesh rule)} \\ i(t) + i_1(t) - C\frac{dv}{dt} = 0 \text{ (node rule)} \end{cases}$$

Intuitively, we can understand that the memory of the system corresponds to the capacitor charge and the electromagnetic flow in the coil. Indeed, if these values are known at time $t = 0$, for a known input, the future of the system is determined in a unique manner. Thus, the possible state variables are given by the values $i(t)$ (proportional to the flow) and $v(t)$ (proportional to the charge). We obtain the state equations by removing the i_1 in the previous equations and by isolating $\frac{di}{dt}$ and $\frac{dv}{dt}$. Of course, one equation must be removed. We obtain:

$$\begin{cases} \frac{dv}{dt} = -\frac{1}{CR_1}v(t) + \frac{1}{C}i(t) + \frac{1}{CR_1}u(t) \\ \frac{di}{dt} = -\frac{1}{L}v(t) - \frac{R_2}{L}i(t) + \frac{1}{L}u(t) \end{cases}$$

Note that the output is given by $y(t) = R_2 i(t)$. Finally, we reach the state representation of a linear system given by:

$$\frac{d}{dt}\begin{pmatrix} v(t) \\ i(t) \end{pmatrix} = \begin{pmatrix} -\frac{1}{CR_1} & \frac{1}{C} \\ -\frac{1}{L} & -\frac{R_2}{L} \end{pmatrix} \begin{pmatrix} v(t) \\ i(t) \end{pmatrix} + \begin{pmatrix} \frac{1}{CR_1} \\ \frac{1}{L} \end{pmatrix} u(t)$$

$$y(t) = \begin{pmatrix} 0 & R_2 \end{pmatrix} \begin{pmatrix} v(t) \\ i(t) \end{pmatrix}$$

Solution to Exercise 1.16 (the three containers)

1) With the aim of applying Bernoulli's relation of the left container, let us consider a flow tube, in other words a virtual tube (see figure) in which the water has a fluid movement and does not cross the walls. Bernoulli's relation tells us that in this tube, at every point:

$$P + \rho \frac{v^2}{2} + \rho g z = \text{constant}$$

where P is the pressure at the considered point, z its height and v the velocity of the water at this point. The coefficient ρ is the bulk density of the water and g is the gravitational constant. Following Bernouilli's relation, we have:

$$P_D + \rho \frac{v_D^2}{2} + \rho g z_D = P_A + \rho \frac{v_A^2}{2} + \rho g z_A$$

in other words:

$$P_D = P_A + \rho g \left(z_A - z_D \right) - \rho \frac{v_D^2}{2}$$

Moreover, we may assume that C is far from the turbulence zone and that the water is not moving. Therefore, we have the following Bernoulli relation:

$$P_C + \rho g z_C = P_B + \rho g z_B$$

in other words:

$$P_C = P_B + \rho g (z_B - z_C)$$

Let us note that, in this turbulence zone, the water is slowed down, but we can assume that the pressure does not change, i.e. $P_C = P_D$. Thus, we have:

$$\underbrace{P_B}_{P_{atm}} + \rho g (z_B - z_C) = \underbrace{P_A}_{P_{atm}} + \rho g (z_A - z_D) - \rho \frac{v_D^2}{2}$$

As $P_A = P_B = P_{atm}$, and that $z_C = z_D$, this equation becomes:

$$\rho g (z_A - z_B) = \rho \frac{v_D^2}{2}$$

or:

$$v_D = \sqrt{2g(z_A - z_B)}$$

In the case where the level of the right container is higher than that of the left one, a similar study gives us:

$$v_D = -\sqrt{2g(z_B - z_A)}$$

The minus sign of the expression indicates that the flow now moves from the right container toward the left one. Thus the general relation for the velocity of the water in the canal is:

$$v_D = sign(z_A - z_B) \sqrt{2g|z_A - z_B|}$$

If a is the cross section of the canal, the water flow from the right container to the left one is:

$$Q_D = a.sign(z_A - z_B) \sqrt{2g|z_A - z_B|}$$

This is the so-called *Torricelli* law.

REMARK.– Initially, this law was proven in a simpler context where the water flows into emptiness. The total energy of a fluid element of mass m is conserved if we consider this latter to be falling freely in the tube flow. Thus, for the two points A and B, where A is at the surface and B in the tube, we have:

$$mgh_A + \underbrace{\frac{1}{2}mv_A^2}_{=0} = mgh_B + \frac{1}{2}mv_B^2$$

and therefore:

$$v_B = \sqrt{2g(h_A - h_B)}$$

We can then deduce Toricelli's relation. This reasoning is only possible in the case of a perfect fluid, where, given the absence of friction, we assume that the forces of tangential pressure are nil.

2) The state equations are obtained by writing that the volume of water in a container is equal to the sum of the incoming flows minus the sum of the outgoing flows, in other words:

$$\dot{h}_1 = -Q_{1\text{ext}} - Q_{12} + u_1$$
$$\dot{h}_2 = Q_{12} - Q_{23}$$
$$\dot{h}_3 = -Q_{3\text{ext}} + Q_{23} + u_2$$

or:

$$\dot{h}_1 = -a.\sqrt{2gh_1} - a.sign(h_1 - h_2)\sqrt{2g|h_1 - h_2|} + u_1$$
$$\dot{h}_2 = a.sign(h_1 - h_2)\sqrt{2g|h_1 - h_2|}$$
$$\quad - a.sign(h_2 - h_3)\sqrt{2g|h_2 - h_3|}$$
$$\dot{h}_3 = -a.\sqrt{2gh_3} + a.sign(h_2 - h_3)\sqrt{2g|h_2 - h_3|} + u_2$$

Solution to Exercise 1.17 (pneumatic cylinder)

The input of the system is the volumetric flow rate u of the air toward the cylinder chamber. We then have:

$$u = \left(\frac{V}{n}\right)\dot{n}$$

where n is the number of gas in the chamber and V is the volume of the chamber. The fundamental principle of dynamics gives us $pa - kz = m\ddot{z}$. Therefore, the first two state equations are:

$$\begin{cases} \dot{z} = \dot{z}, \\ \ddot{z} = \frac{ap-kz}{m} \end{cases}$$

The ideal gas law ($pV = nRT$) is given by $pza = nRT$. By differentiating, we obtain:

$$a\left(\dot{p}z + p\dot{z}\right) = R\left(\dot{n}T + n\dot{T}\right)$$

By assuming an isothermal evolution, this relation becomes:

$$a\left(\dot{p}z + p\dot{z}\right) = R\dot{n}T = R\frac{nu}{V}T = pu$$

By isolating $\dot{p}$, we obtain the third state equation of our system, which is:

$$\dot{p} = \frac{p}{z}\left(\frac{u}{a} - \dot{z}\right)$$

The state equations of the system are therefore:

$$\begin{cases} \dot{z} = \dot{z} \\ \ddot{z} = \frac{ap-kz}{m} \\ \dot{p} = \frac{p}{z}\left(\frac{u}{a} - \dot{z}\right) \end{cases}$$

or, since $\mathbf{x} = (z, \dot{z}, p)$:

$$\begin{cases} \dot{x}_1 = x_2 \\ \dot{x}_2 = \frac{ax_3 - kx_1}{m} \\ \dot{x}_3 = -\frac{x_3}{x_1}\left(x_2 - \frac{u}{a}\right) \end{cases}$$

Solution to Exercise 1.18 (Fibonacci sequence)

1) The state equations are given by:

$$\begin{cases} x_1(k+1) = x_2(k) \\ x_2(k+1) = x_1(k) + x_2(k) \\ y(k) = x_1(k) + x_2(k) \end{cases}$$

where $x_1(0) = 1$ and $x_2(0) = 0$ as the initial conditions. This system is called a *Fibonacci* system.

2) Let us now look for the recurrence relation associated with this system. For this we need to express $y(k)$, $y(k+1)$ and $y(k+2)$ as a function of $x_1(k)$ and $x_2(k)$. The resulting calculations are the following:

$$y(k) = x_1(k) + x_2(k)$$
$$y(k+1) = x_1(k+1) + x_2(k+1) = x_1(k) + 2x_2(k)$$
$$y(k+2) = x_1(k+2) + x_2(k+2) = x_1(k+1) + 2x_2(k+1)$$
$$= 2x_1(k) + 3x_2(k)$$

In other words:

$$\begin{pmatrix} y(k) \\ y(k+1) \\ y(k+2) \end{pmatrix} = \begin{pmatrix} 1 & 1 \\ 1 & 2 \\ 2 & 3 \end{pmatrix} \begin{pmatrix} x_1(k) \\ x_2(k) \end{pmatrix}$$

By removing $x_1(k)$ and $x_2(k)$ from this system of three linear equations, we obtain a single equation given by:

$$y(k+2) - y(k+1) - y(k) = 0$$

The initial conditions are $y(0) = y(1) = 1$. It is in general in this form that the Fibonacci system is described.

Solution to Exercise 1.19 (bus network)

1) The timetable is given below.

k	1	2	3	4	5
$x_1(k)$	0	5	8	13	16
$x_2(k)$	0	3	8	11	16

Table 1.2. *Timetable for the buses*

We note a periodicity of 2 in the sequence progression.

2) The state equations are:

$$\begin{cases} x_1(k+1) = \max(x_1(k) + 2, x_2(k) + 5) \\ x_2(k+1) = \max(x_1(k) + 3, x_2(k) + 3) \end{cases}$$

3) In matrix form, we have:

$$\mathbf{x}(k+1) = \begin{pmatrix} 2 & 5 \\ 3 & 3 \end{pmatrix} \otimes \mathbf{x}(k)$$

4) The problem comes from the fact that $(\mathbb{R}, \oplus)$ is only a monoid and not a group (since the image does not exist). Therefore $(\mathbb{R}, \oplus, \otimes)$ is not a ring (as it is not necessary for matrix calculus) but a dioid.

2

Simulation

In this chapter, we will show how to perform a computer simulation of a nonlinear system described by its state equations:

$$\begin{cases} \dot{\mathbf{x}}(t) = \mathbf{f}(\mathbf{x}(t), \mathbf{u}(t)) \\ \mathbf{y}(t) = \mathbf{g}(\mathbf{x}(t), \mathbf{u}(t)) \end{cases}$$

This step is important in order to test the behavior of a system (controlled or not). Before presenting the simulation method, we will introduce the concept of vector fields. This concept will allow us to better understand the simulation method as well as certain behaviors that could appear in nonlinear systems. We will also give several concepts of graphics that are necessary for the graphical representation of our systems.

2.1. Concept of vector field

We will now present the concept of vector fields and show the manner in which they are useful in order to better understand the various behaviors of systems. We invite the readers to consult Khalil [KHA 02] for further details on this subject. A *vector field* is a continuous function f of $\mathbb{R}^n$ to $\mathbb{R}^n$.

When $n = 2$, a graphical representation of the function f can be imagined. For instance, the vector field associated with the linear function:

$$\mathbf{f} : \begin{matrix} \mathbb{R}^2 \to \mathbb{R}^2 \\ \begin{pmatrix} x_1 \\ x_2 \end{pmatrix} \to \begin{pmatrix} x_1 + x_2 \\ x_1 - x_2 \end{pmatrix} \end{matrix}$$

is illustrated in Figure 2.1. In order to obtain this figure, we have taken a set of vectors from the initial set, following a grid. Then, for each grid vector x, we have drawn its image vector $\mathbf{f}(\mathbf{x})$ by giving it the vector x as origin.

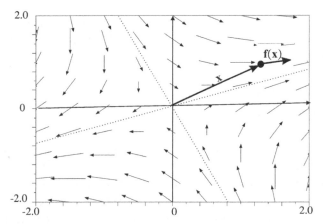

Figure 2.1. *Vector field associated with a linear application*

The MATLAB code that allowed us to generate such a field is given below:

```
axis([-2,2,-2,2]);
Mx = -2:0.5:2; My = -2:0.5:2;
[X1,X2] = meshgrid(Mx,My);
VX=X1+X2; VY=X1-X2;
quiver(Mx,My,VX,VY,'black');
```

This program can also be found in the file `field_syslin.m`. We may recognize in this figure the characteristic spaces (dotted lines) of the linear application. We can also see that one eigenvalue is positive and another eigenvalue is negative. This can be verified by analyzing the matrix of our linear application given by:

$$\mathbf{A} = \begin{pmatrix} 1 & 1 \\ 1 & -1 \end{pmatrix}$$

Its eigenvalues are $\sqrt{2}$ and $-\sqrt{2}$, and the associated eigenvectors are:

$$\mathbf{v}_1 = \begin{pmatrix} 0.9239 \\ 0.3827 \end{pmatrix} \text{ et } \mathbf{v}_2 = \begin{pmatrix} -0.3827 \\ 0.9239 \end{pmatrix}$$

Let us note that the vector x represented in the figure is not an eigenvector, since x and f(x) are not collinear. However, all the vectors that belong to the characteristic subspaces (represented as dotted lines in the figure) are eigenvectors. Along the characteristic subspace associated with the negative eigenvalue, the field vectors tend to point toward 0, whereas these vectors point to infinity along the characteristic subspace associated with the positive eigenvalue.

For an autonomous system (in other words, one without input), the evolution is given by the equation $\dot{\mathbf{x}}(t) = \mathbf{f}(\mathbf{x}(t))$. When f is a function of $\mathbb{R}^2$ to $\mathbb{R}^2$, we can obtain a graphical representation of f by drawing the vector field associated with f. The graph will then allow us to better understand the behavior of our system.

2.2. Graphical representation

In this section, we will give several concepts that are necessary for the graphical representation of systems during simulations.

2.2.1. *Patterns*

A *pattern* is a matrix with two or three rows (following whether the object is in the plane or in space) and n columns that represent the n vertices of a shape-retaining polygon, meant to represent the object. It is important that the unions of all the segments formed by the two consecutive points of the pattern form the edges of the polygon that we wish to represent. For instance, the pattern M of the chassis (see Figure 2.2) of the car (with the rear wheels) is given by:

$$\begin{pmatrix} -1 & 4 & 5 & 5 & 4 & -1 & -1 & 0 & 0 & -1 & 1 & 0 & 0 & -1 & 1 & 0 & 0 & 3 & 3 & 3 \\ -2 & -2 & -1 & 1 & 2 & 2 & -2 & -2 & -2 & -3 & -3 & -3 & -3 & 3 & 3 & 3 & 3 & 2 & 2 & 3 & -3 \end{pmatrix}$$

It is clear on the graph of the car in movement that the front wheels can move with respect to the chassis, as well as with respect to one another. They, therefore, cannot be incorporated into the pattern of the chassis. For the graph of the car, we will, therefore, have to use 3 patterns: that of the chassis, that of the left front wheel and that of the right front wheel. In MATLAB, the pattern M (here in two dimensions) can be drawn in blue using the instructions:

```
M=[-1 4 5 5 4 -1 -1 -1 0 0 -1 1 0 0 -1 1 0 0 3 3 3;
-2 -2 -1 1 2 2 -2 -2 -2 -3 -3 -3 -3 3 3 3 3 2 2 3 -3];
plot(M(1, :),M(2, :),'blue');
```

Recall that in MATLAB, M(i, :) returns the ith row of the matrix M.

2.2.2. *Rotation matrix*

Let us recall that the jth column of the matrix of a linear application of $\mathbb{R}^n \to \mathbb{R}^n$ represents the image of the jth vector

e_j of the standard basis. Thus, the expression of a rotation matrix of angle θ in the plane $\mathbb{R}^2$ is given by (see Figure 2.3):

$$\mathbf{R} = \begin{pmatrix} \cos\theta & -\sin\theta \\ \sin\theta & \cos\theta \end{pmatrix}$$

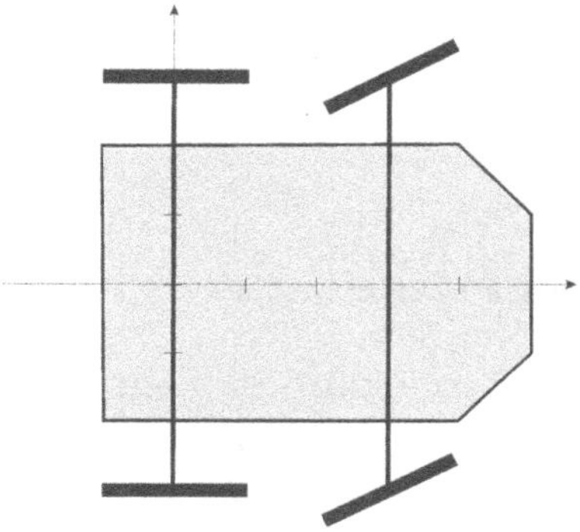

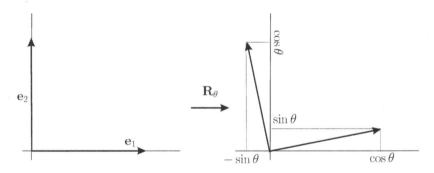

Figure 2.2. *Car to be represented graphically*

Figure 2.3. *Rotation of angle θ*

Concerning rotations in the space $\mathbb{R}^3$, it is important to specify the rotation axis. We can distinguish between 3 main rotations: the rotation of angle θ around the Ox axis, the rotation around the Oy axis and the rotation around the Oz

axis. The associated matrices are, respectively, given by:

$$\mathbf{R}_x = \begin{pmatrix} 1 & 0 & 0 \\ 0 & \cos\theta & -\sin\theta \\ 0 & \sin\theta & \cos\theta \end{pmatrix}, \mathbf{R}_y = \begin{pmatrix} \cos\theta & 0 & \sin\theta \\ 0 & 1 & 0 \\ -\sin\theta & 0 & \cos\theta \end{pmatrix}$$

$$\text{and } \mathbf{R}_z = \begin{pmatrix} \cos\theta & -\sin\theta & 0 \\ \sin\theta & \cos\theta & 0 \\ 0 & 0 & 1 \end{pmatrix}$$

2.2.3. *Homogeneous coordinates*

Drawing two-dimensional or three-dimensional (3D) objects on a screen requires a series of affine transformations (rotations, translations and homotheties) of the form:

$$\mathbf{f}_i : \begin{array}{l} \mathbb{R}^n \to \mathbb{R}^n \\ \mathbf{x} \mapsto \mathbf{A}_i \mathbf{x} + \mathbf{b}_i \end{array}$$

with $n = 2$ or 3. However, the manipulation of compositions of affine functions is not as simple as that of linear applications. The idea of the transformation into *homogeneous coordinates* is to transform a system of affine equations into a system of linear equations. Let us note first of all that an affine equation of type $\mathbf{y} = \mathbf{Ax} + \mathbf{b}$ can be rewritten in the form:

$$\begin{pmatrix} \mathbf{y} \\ 1 \end{pmatrix} = \begin{pmatrix} \mathbf{A} & \mathbf{b} \\ 0 & 1 \end{pmatrix} \begin{pmatrix} \mathbf{x} \\ 1 \end{pmatrix}$$

We will, therefore, define the *homogeneous transformation* of a vector as follows:

$$\mathbf{x} \mapsto \mathbf{x}_h = \begin{pmatrix} \mathbf{x} \\ 1 \end{pmatrix}$$

Thus, an equation such as:

$$y = A_3 \left(A_2 \left(A_1 x + b_1 \right) + b_2 \right) + b_3$$

where there is a composition of 3 affine transformations can be rewritten as:

$$y_h = \begin{pmatrix} A_3 & b_3 \\ 0 & 1 \end{pmatrix} \begin{pmatrix} A_2 & b_2 \\ 0 & 1 \end{pmatrix} \begin{pmatrix} A_1 & b_1 \\ 0 & 1 \end{pmatrix} x_h$$

By using *Rodrigues' formula* that tells us that the rotation matrix around the vector w and of angle $\varphi = ||w||$ is given by:

$$R_w = \exp \begin{pmatrix} 0 & -w_z & w_y \\ w_z & 0 & -w_x \\ -w_y & w_x & 0 \end{pmatrix}$$

we can write a MATLAB function to generate a homogeneous rotation matrix of $\mathbb{R}^3$:

```
function R=Rotate(w)
  A=[ 0 -w(3) w(2); w(3) 0 -w(1); -w(2) w(1) 0]);
  R=[expm(A),[0;0;0]; 0 0 0 1];
```

A function that generates a homogeneous translation matrix of a vector v of $\mathbb{R}^3$ is given below:

```
function T=Translate(v)
  T=eye(4,4);
  T(1:3,4)=v;
```

These two functions are given in the files Rotate.m and Translate.m.

2.3. Simulation

In this section, we will present the integration method to perform a computer simulation of a nonlinear system described by its state equations:

$$\begin{cases} \dot{\mathbf{x}}(t) = \mathbf{f}(\mathbf{x}(t), \mathbf{u}(t)) \\ \mathbf{y}(t) = \mathbf{g}(\mathbf{x}(t), \mathbf{u}(t)) \end{cases}$$

This method is rather approximative, but remains simple to understand and is enough in order to describe the behaviors of most of the robotized systems.

2.3.1. *Euler's method*

Let dt be a very small number compared to the time constants of the system and which corresponds to the sampling period of the method (for example, $dt = 0.01$). The evolution equation is approximated by:

$$\frac{\mathbf{x}(t+dt) - \mathbf{x}(t)}{dt} \simeq \mathbf{f}(\mathbf{x}(t), \mathbf{u}(t))$$

in other words:

$$\mathbf{x}(t+dt) \simeq \mathbf{x}(t) + \mathbf{f}(\mathbf{x}(t), \mathbf{u}(t)).dt$$

This equation can be interpreted as an order 1 Taylor formula. From this, we can deduce the simulation algorithm (called *Euler's method*):

Algorithm	EULER(in: $\mathbf{x}_0$)
1	$\mathbf{x} := \mathbf{x}_0; t := 0;\ dt = 0.01;$
2	repeat
3	wait for uinput;
4	$\mathbf{y} := \mathbf{g}(\mathbf{x}, \mathbf{u});$
5	return $\mathbf{y}$;
6	$\mathbf{x} := \mathbf{x} + \mathbf{f}(\mathbf{x}, \mathbf{u}).dt;$
7	wait for interrupt from timer;
8	$t = t + dt;$
9	while true

The timer creates a periodic interrupt every dt s. Thus, if the computer is sufficiently fast, the simulation is performed at the same speed as our physical system. We then refer to *real-time* simulation. In some circumstances, what we are interested in is obtaining the result of the simulation in the fastest possible time (for instance, in order to predict how a system will behave in the future). In this case, it is not necessary to slow the computer down in order to synchronize it with our physical time.

We call *local error* the quantity:

$$e_t = ||\mathbf{x}(t+dt) - \hat{\mathbf{x}}(t+dt)|| \text{ with } \mathbf{x}(t) = \hat{\mathbf{x}}(t)$$

where $\mathbf{x}(t+dt)$ is the exact solution of the differential equation $\dot{\mathbf{x}} = \mathbf{f}(\mathbf{x}, \mathbf{u})$ and $\hat{\mathbf{x}}(t+dt)$ is the estimated value of the state vector, for the integration scheme being used. For Euler's method, we can show that e_t is of order 1, i.e. $e_t = o(dt)$.

2.3.2. *Runge–Kutta method*

There are more efficient integration methods in which the local error is of order 2 or more. This is the case of the Runge–Kutta method of order 2, which consists of replacing the recurrence $\hat{\mathbf{x}}(t+dt) := \hat{\mathbf{x}}(t) + \mathbf{f}(\hat{\mathbf{x}}(t), \mathbf{u}(t)).dt$ with:

$$\hat{\mathbf{x}}(t+dt) = \hat{\mathbf{x}}(t) + dt.\left[\tfrac{1}{4}.\mathbf{f}(\hat{\mathbf{x}}(t), \mathbf{u}(t)) \right.$$
$$\left. + \tfrac{3}{4}.\mathbf{f}(\underbrace{\hat{\mathbf{x}}(t) + \tfrac{2}{3}dt.\mathbf{f}(\hat{\mathbf{x}}(t), \mathbf{u}(t))}_{\hat{\mathbf{x}}_E(t+\tfrac{2}{3}dt)}, \mathbf{u}(t+\tfrac{2}{3}dt)) \right]$$

Let us note in this expression the value $\hat{\mathbf{x}}_E(t+\tfrac{2}{3}dt)$ which can be interpreted as the integration obtained by Euler's method at time $t + \tfrac{2}{3}dt$. The quantity between the square brackets is an average between an estimation of $\mathbf{f}(\mathbf{x}(t),\mathbf{u}(t))$ and an estimation of $\mathbf{f}\left(\hat{\mathbf{x}}(t+\frac{2}{3}dt),\mathbf{u}(t+\frac{2}{3}dt)\right)$. The local error e_t here is of order 2 and the integration method is, therefore, a lot more precise. There are Runge–Kutta methods of order higher than 2 that we will not discuss here.

2.3.3. *Taylor's method*

Euler's method (which is an order 1 Taylor method) can be extended to higher orders. Let us show, without loss of generality, how to extend to second order. We have:

$$\mathbf{x}(t+dt) = \mathbf{x}(t) + \dot{\mathbf{x}}(t).dt + \ddot{\mathbf{x}}(t).dt^2 + o\left(dt^2\right)$$

But:

$$\dot{\mathbf{x}}(t) = \mathbf{f}(\mathbf{x}(t),\mathbf{u}(t))$$
$$\ddot{\mathbf{x}}(t) = \frac{\partial \mathbf{f}}{\partial \mathbf{x}}(\mathbf{x}(t),\mathbf{u}(t)).\dot{\mathbf{x}}(t) + \frac{\partial \mathbf{f}}{\partial \mathbf{u}}(\mathbf{x}(t),\mathbf{u}(t)).\dot{\mathbf{u}}(t)$$

Therefore, the integration scheme becomes:

$$\hat{\mathbf{x}}(t+dt) = \hat{\mathbf{x}}(t) + dt.\mathbf{f}(\hat{\mathbf{x}}(t),\mathbf{u}(t)) + dt^2$$
$$\cdot \left(\frac{\partial \mathbf{f}}{\partial \mathbf{x}}(\hat{\mathbf{x}}(t),\mathbf{u}(t)).\mathbf{f}(\hat{\mathbf{x}}(t),\mathbf{u}(t)) + \frac{\partial \mathbf{f}}{\partial \mathbf{u}}(\hat{\mathbf{x}}(t),\mathbf{u}(t)).\dot{\mathbf{u}}(t) \right)$$

2.4. Exercises

EXERCISE 2.1.– Vector field of the predator–prey system

The *predator–prey* system, also called a Lotka–Volterra system, is given by:

$$\begin{cases} \dot{x}_1(t) = (1 - x_2(t))\, x_1(t) \\ \dot{x}_2(t) = (x_1(t) - 1)\, x_2(t) \end{cases}$$

The state variables $x_1(t)$ and $x_2(t)$ represent the size of the predator and prey populations. For example, x_1 could represent the number of preys in thousands, whereas x_2 could be the number of predators in thousands. Even though the number of preys and predators are integers, we will assume that x_1 and x_2 are real values. The quadratic terms of this state equation represent the interactions between the two species. Let us note that the preys grow in an exponential manner when there are no predators. Similarly, the population of predators declines when there is no prey.

1) Figure 2.4 corresponds to the vector field associated with the evolution function:

$$\mathbf{f}(\mathbf{x}) = \begin{pmatrix} (1 - x_2)\, x_1 \\ (x_1 - 1)\, x_2 \end{pmatrix}$$

on the grid $[0, 2] \times [0, 2]$. Discuss the dynamic behavior of the system using this figure.

2) Also using this figure, give the point of equilibrium. Verify it through calculation.

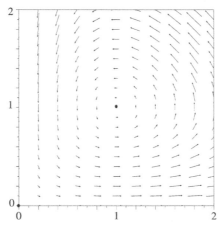

Figure 2.4. *Vector field associated with the Lotka–Volterra system, in the plane (x_1, x_2)*

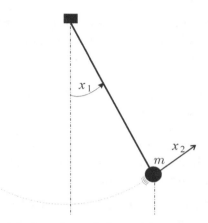

Figure 2.5. *Simple pendulum with state vector*
$$\mathbf{x} = (x_1, x_2)^T = \left(\theta, \dot{\theta}\right)^T$$

EXERCISE 2.2.– Vector field of a simple pendulum

Let us consider the simple pendulum described by the following state equations:

$$\begin{cases} \dot{x}_1 = x_2 \\ \dot{x}_2 = -g \sin x_1 \end{cases}$$

The vector field associated with the evolution function $\mathbf{f}(\mathbf{x})$ is drawn in Figure 2.6.

1) Following the graph, give the stable and unstable points of equilibrium.

2) Draw on the figure, a path for the pendulum that could have been obtained by Euler's method.

EXERCISE 2.3.– Pattern of a cube

Let us consider the 3D cube $[0, 1] \times [0, 1] \times [0, 1]$.

1) Give its pattern in matrix form.

2) What matrix operation must be performed in order to rotate it with an angle θ around the Ox axis?

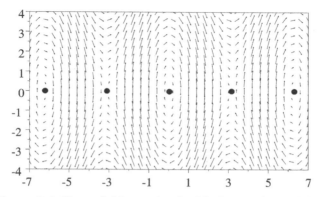

Figure 2.6. *Vector field associated with the simple pendulum in the phase plane* $(x_1, x_2) = \left(\theta, \dot{\theta}\right)$

EXERCISE 2.4.– Drawing a car

Here, we are looking to design a MATLAB function that we will call voiture_draw(x), which draws a car in a state $\mathbf{x} = (x, y, \theta, v, \delta)^T$ where x, y, θ correspond to the pose of the car (in other words, its position and orientation), v is the speed and δ is the angle of the front wheels.

1) Define, in homogeneous coordinates, a pattern for the chassis and a common pattern for the two front wheels.

2) Define the transformation matrices for the chassis, left front wheel and right front wheel.

3) Deduce from this, a MATLAB function called car_draw(x) that draws the car in a given state x.

EXERCISE 2.5.– Simulation of a pendulum

Let us consider a pendulum described by the following differential equation:

$$\ddot{\theta} = \frac{-g \sin \theta}{\ell}$$

where θ represents the angle of the pendulum. We will take $g = 9.81$ ms^{-2} and $\ell = 1$ m. We initialize the pendulum at time $t = 0$ with $\theta = 1$ rad and $\dot{\theta} = 0$ rad.s^{-1}. Then, we let go of the pendulum. Write a small program (influenced by a MATLAB-like syntax) that determines the angle of the pendulum at time $t = 1$ s. The program will have to use Euler's method.

EXERCISE 2.6.– Van der Pol system

Let us consider the system described by the following differential equation:

$$\ddot{y} + (y^2 - 1)\dot{y} + y = 0$$

1) Let us choose as state vector $\mathbf{x} = (y \ \dot{y})^{\mathrm{T}}$. Give the state equations of the system.

2) Linearize this system around the point of equilibrium. What are the poles of the system? Is the system stable around the equilibrium point?

3) The vector field associated with this system is represented in Figure 2.7 in the state space (x_1, x_2). We initialize the system in $\mathbf{x}_0 = (0.1 \ 0)^{\mathrm{T}}$. Draw on the figure, the path $\mathbf{x}(t)$ of the system on the state space. Give the form of $y(t)$.

4) Can a path have a loop?

5) In order to simulate this system, would it be better to employ Euler's method, the Runge–Kutta method, or will both integration schemes give equivalent behaviors?

EXERCISE 2.7.– Simulation of a car

Let us consider the car with the following state equations:

$$\begin{pmatrix} \dot{x} \\ \dot{y} \\ \dot{\theta} \\ \dot{v} \\ \dot{\delta} \end{pmatrix} = \begin{pmatrix} v \cos \delta \cos \theta \\ v \cos \delta \sin \theta \\ \frac{v \sin \delta}{L} \\ u_1 \\ u_2 \end{pmatrix}$$

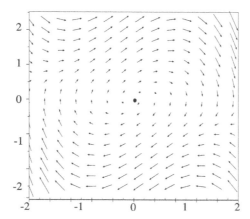

Figure 2.7. *Vector field associated with the Van der Pol system*

The state vector is given by $\mathbf{x} = (x, y, \theta, v, \delta)^\mathrm{T}$. Let us take the initial state for the car as $\mathbf{x}(0) = (0\ 0\ 0\ 7\ 0)^\mathrm{T}$, which means that at time $t = 0$, the car is centered around the origin, with a nil heading angle, a speed of 7 ms^{-1} and the front wheels parallel to the axis of the car. We assume that the vectorial control $\mathbf{u}(t)$ remains constant and equal to $(0\ 0.2)^\mathrm{T}$. Which means that the car does not accelerate (since $u_1 = 0$) and that the steering wheel is turning at a constant speed of $0.2\ rad.s^{-1}$ (since $u_2 = 0.2$). Give a MATLAB program that simulates the dynamic evolution of this car during 10 s with Euler's method and a sampling step of 0.01 s.

EXERCISE 2.8.– Integration by Taylor's method

Let us consider a robot (such as, a tank) described by the following state equations:

$$\begin{pmatrix} \dot{x}_1 \\ \dot{x}_2 \\ \dot{x}_3 \\ \dot{x}_4 \\ \dot{x}_5 \end{pmatrix} = \begin{pmatrix} x_4 \cos x_3 \\ x_4 \sin x_3 \\ x_5 \\ u_1 \\ u_2 \end{pmatrix}$$

Propose a second-order integration scheme using Taylor's method. The input will be:

$$\mathbf{u} = \begin{pmatrix} u_1 \\ u_2 \end{pmatrix} = \begin{pmatrix} \cos(t) \\ \sin(t) \end{pmatrix}$$

EXERCISE 2.9.– Radius of a tricycle wheel

Let us consider the 3D representation of a tricycle as shown in Figure 2.8. In this figure, the small black disks are on a same horizontal plane of height r and the small hollow disks are on the same horizontal plane with height nil.

The radius of the front wheel in bold has an angle of α with the horizontal plane, as represented in the figure. Give, as a function of x, y, θ, δ, α, L, r, the expression of the transformation matrix (in homogeneous coordinates) that links the radius positioned on the Ox axis with that (in bold) of the front wheel. The expression must have a matrix product form.

EXERCISE 2.10.– Three-dimensional simulation of a tricycle in MATLAB

The driver of the tricycle of Figure 2.9 has two controls: the acceleration of the front wheel and the rotation speed of the steering wheel. The state variables of our system are composed of the position coordinates (the x, y coordinates of the center if the rear axle, the orientation θ of the tricycle and the angle δ of the front wheel) and of the speed v of the center of the front wheel. As seen in Exercise 1.11, the evolution equation of the tricycle (similar to that of the car) is written as:

$$\begin{pmatrix} \dot{x} \\ \dot{y} \\ \dot{\theta} \\ \dot{v} \\ \dot{\delta} \end{pmatrix} = \begin{pmatrix} v \cos \delta \cos \theta \\ v \cos \delta \sin \theta \\ \frac{v \sin \delta}{L} \\ u_1 \\ u_2 \end{pmatrix}$$

where $L = 3$ is the distance between the rear axle and the center of the front wheel. The state vector here is $\mathbf{x} = (x, y, \theta, v, \delta)^{\mathrm{T}}$. The distance between each rear wheel and the axis of the tricycle is given by $e = 2$ m. The radius of the wheels is $r = 1$ m.

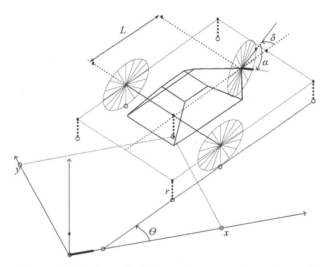

Figure 2.8. *We are looking to draw a radius of one of the tricycle wheels*

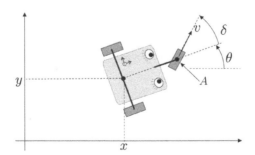

Figure 2.9. *Robotized tricycle to be simulated*

1) Create a pattern for the body of the tricycle. Rotate and translate it in 3D using the homogeneous coordinates and Rodrigues' formula.

2) Create a function `tricycle_draw` that draws this tricycle in three dimensions together with the wheels, as in Figure 2.10. This function will have as parameter the state vector x.

3) By using the speed composition rule, calculate the speeds that each wheel must have.

4) In order to rotate the wheels (numbered from 1 to 3 as in the figure) with the progress of the tricycle, we add 3 state variables α_1, α_2, α_3 corresponding to the angle of the wheels. Give the new state equations of the system. Simulate the system in MATLAB.

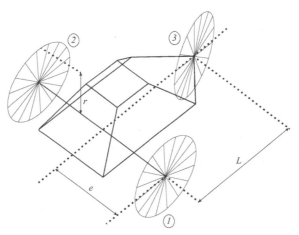

Figure 2.10. *Tricycle to be drawn*

EXERCISE 2.11.– Manipulator arms

The manipulator robot represented in Figure 2.11 is composed of three arms in series. The first arm, of length 3, can rotate around the Oz axis. The second arm, of length 2, placed at the end of the first arm can also rotate around the Oz axis. The third arm, of length 1, placed at the end of the second arm, can rotate around the axis formed by the second arm. This robot has 3 degrees of freedom $x = (\alpha_1, \alpha_2, \alpha_3)$, where the α_i represents the angles formed by each of the arms. The basic pattern chosen to represent each of the arms is the unit cube. Each arm is assumed to be a parallelepiped of thickness 0.3. In order to take the form of the arm, the pattern must be subjected to an affinity, represented by a diagonal matrix. Then, it has to be rotated and translated in order to be correctly placed. Design a MATLAB program that simulates this system, with a 3D vision inspired from the figure.

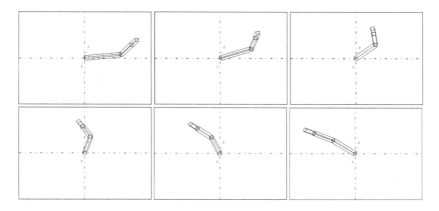

Figure 2.11. *Manipulator robot composed of three arms*

EXERCISE 2.12.– Simulation and control of a skating robot in MATLAB

The skating vehicle represented in Figure 2.12 is designed to move on a frozen lake and stands on five ice-skates [JAU 10]. This system has two inputs: the tangent u_1 of angle β of the front skate (we have chosen the tangent as input in order to avoid singularities) and u_2 the torque exerted on the articulation between the two sledges and which corresponds to the angle δ. The propulsion, therefore, only comes from the torque u_2 and is similar to the mode of propulsion of a snake or an eel [BOY 06]. Any control over u_1, therefore, does not give any energy to the system.

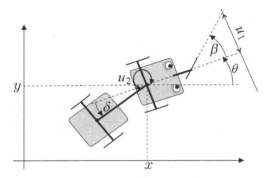

Figure 2.12. *Skating robot*

1) Show that the system can be described by the following state equations:

$$\begin{cases} \dot{x} = v\cos\theta \\ \dot{y} = v\sin\theta \\ \dot{\theta} = vu_1 \\ \dot{v} = -\left(u_1 + \sin\delta\right)u_2 - v \\ \dot{\delta} = -v\left(u_1 + \sin\delta\right) \end{cases}$$

where v is the speed of the center of the front axle. This is of course a simplified and normalized model where, for reasons of simplicity, the coefficients (masses, viscous friction, interaxle distances, etc.) have been chosen unitary.

2) Simulate this system by using Euler's method for 2 s. We will take as initial condition $\mathbf{x} = (0, 0, 0, 3, 0.5)^T$, as sampling period $dt = 0.05$ s and as input $\mathbf{u} = (1, 0)^T$. Discuss the result.

3) We will now attempt to control this system using *biomimicry*, i.e. by attempting to reproduce the propulsion of the snake. We choose u_1 to be of the form:

$$u_1 = p_1 \cos(p_2 t) + p_3$$

where p_1 is the amplitude, p_2 is the pulse and p_3 is the bias. We choose u_2 so that the propulsion torque is always a driving force, in other words $\dot{\delta}.u_2 \geq 0$. Indeed, the term $\dot{\delta}.u_2$ corresponds to the power brought by the robot which is transformed into kinetic energy. Program this control and make the right choice for the parameters that allow us to ensure an efficient propulsion. Reproduce the two behaviors illustrated in Figure 2.13 on your computer.

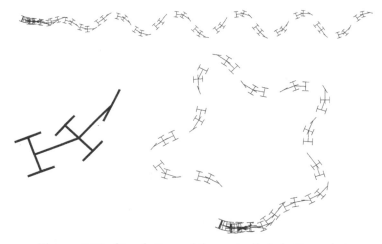

Figure 2.13. *Simulations of the controlled skating robot*

4) Add a second control loop that controls the parameters p_i of your controller in order for your robot to be able to follow a desired heading $\bar{\theta}$.

2.5. Solutions

Solution to Exercise 2.1 (vector field of the predator–prey system)

1) The evolution of the state vector of the system is done in the direction of the arrows (since, $\dot{\mathbf{x}} = \mathbf{f(x)}$). We can thus observe that the evolution of the system is periodic and that the state traverses an almost-circular curve (with a center of $(1,1)$) in the direct trigonometric direction.

2) The point $(1,1)$ and the point $(0,0)$ of the figure (represented by small black disks) appear to be points of equilibrium of our system, since the vector field is canceled out. Let us verify this by calculation. The points of equilibrium satisfy the equation $f(x) = 0$, i.e.:

$$\begin{cases} (1 - \bar{x}_2)\bar{x}_1 = 0 \\ (-1 + \bar{x}_1)\bar{x}_2 = 0 \end{cases}$$

The first point of equilibrium is $\bar{x} = (0, 0)^T$, which corresponds to a situation in which none of the two species exists. The second point of equilibrium is given by $\bar{x} = (1, 1)^T$ and corresponds to a situation of ecological equilibrium.

Solution to Exercise 2.2 (vector field of a simple pendulum)

1) The small black disks represent the points of equilibrium. The first, the third and the last points correspond to the situation where the pendulum is at the bottom, in stable equilibrium. The second and fourth points correspond to the situation where pendulum is on top, in an unstable equilibrium. Around the stable point of equilibrium, the state vector tends to rotate around this point, thus forming a cycle.

2) Figure 2.14 represents the path of the pendulum obtained by Euler's method in the state space for the initial condition $\theta = 1$ and $\dot{\theta} = 0$. The pendulum, which is not subject to any friction, should normally traverse a cycle. However, Euler's method gives a bit of energy to the pendulum at each step, which would explain its path that tends to diverge. This comes from the fact that the field, even though tangent to the path of the system, tends to leave the system. It is clear that Euler's method cannot be used to simulate conservative systems (such as, the planetary system), which do not have friction and tend to perform cycles. However, for dissipative systems (with friction), Euler's method often proves to be sufficient to correctly describe the behavior of the system. For

our pendulum, Euler's method tends to give some energy to the pendulum which, after around 10 oscillations, starts to perform complete revolutions around its axis, in the indirect trigonometric direction.

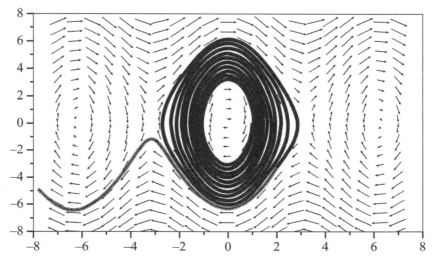

Figure 2.14. *Path of the pendulum in the state space*

Solution to Exercise 2.3 (pattern of a cube)

1) The pattern is given by:

$$\mathbf{M} = \begin{pmatrix} 0 & 1 & 1 & 0 & 0 & 0 & 1 & 1 & 0 & 0 & 0 & 1 & 1 & 1 & 1 \\ 0 & 0 & 0 & 0 & 0 & 1 & 1 & 1 & 1 & 1 & 1 & 0 & 0 & 1 & 1 & 0 \\ 0 & 0 & 1 & 1 & 0 & 0 & 0 & 1 & 1 & 0 & 1 & 1 & 1 & 1 & 0 & 0 \end{pmatrix}$$

is associated with the unit cube $[0, 1]^3$ of $\mathbb{R}^3$. Let us note that this pattern is composed of 16 columns, instead of 13 (indeed, the cube has 12 edges). This comes from the fact that, in order to draw all the edges of the cube without lifting the pen, we necessarily need to go through a minimum of 13 vertices of the cube.

2) We perform the following matrix operation:

$$\mathbf{M} = \begin{pmatrix} 1 & 0 & 0 \\ 0 & \cos\theta & -\sin\theta \\ 0 & \sin\theta & \cos\theta \end{pmatrix} \cdot \mathbf{M}$$

Solution to Exercise 2.4 (drawing a car)

1) For the chassis, we can take the pattern $\mathbf{M}_{\text{chassis}}$ given by:

$$\begin{pmatrix} -1 & 4 & 5 & 5 & 4 & -1 & -1 & 0 & 0 & -1 & 1 & 0 & 0 & -1 & 1 & 0 & 0 & 3 & 3 & 3 \\ -2 & -2 & -1 & 1 & 2 & 2 & -2 & -2 & -3 & -3 & -3 & -3 & 3 & 3 & 3 & 3 & 2 & 2 & 3 & -3 \\ 1 & 1 & 1 & 1 & 1 & 1 & 1 & 1 & 1 & 1 & 1 & 1 & 1 & 1 & 1 & 1 & 1 & 1 & 1 & 1 \end{pmatrix}$$

For this, we have taken the pattern for the chassis of the car, and in order to make it homogeneous, we have added a line of 1s. For the pattern of the left front wheel (as well as the right front wheel), we will take, in homogeneous coordinates, the following pattern:

$$\mathbf{M}_{\text{wheel}} = \begin{pmatrix} -1 & 1 \\ 0 & 0 \\ 1 & 1 \end{pmatrix}$$

2) We have to subject the chassis to a rotation of angle θ and a translation of vector (x, y). For this, it is sufficient to multiply M from the left-hand side by the matrix:

$$\mathbf{R}_{\text{chassis}} = \begin{pmatrix} \cos\theta & -\sin\theta & x \\ \sin\theta & \cos\theta & y \\ 0 & 0 & 1 \end{pmatrix}$$

For the left front wheel, we give it a rotation of angle δ followed by a translation of $(3, 3)$, then rotate it again by θ and translate it by (x, y). The resulting transformation matrix is:

$$\mathbf{R}_{\text{left\_front\_wheel}} = \begin{pmatrix} \cos\theta & -\sin\theta & x \\ \sin\theta & \cos\theta & y \\ 0 & 0 & 1 \end{pmatrix} \begin{pmatrix} \cos\delta & -\sin\delta & 3 \\ \sin\delta & \cos\delta & 3 \\ 0 & 0 & 1 \end{pmatrix}$$

A similar matrix can be obtained for the right front wheel.

3) We define first of all the pattern for the chassis of the car (with its rear wheels) with a pattern for a front wheel (left or right). They are then subjected to the transformations and finally draw the transformed patterns. We obtain the following MATLAB function.

```
function voiture_draw(x)
Mch= [-1 4 5 5 4 -1 -1 -1 0 0 -1 1 0 0 -1 1 0 0 3 3 3;
-2 -2 -1 1 2 2 -2 -2 -2 -3 -3 -3 -3 3 3 3 3 2 2 3 -3;
ones(1,21)]; % Pattern for the chassis
Mav= [-1 1;0 0;1 1]; %Pattern for a front wheel
Rch= [cos(x(3)),-sin(x(3)),x(1);sin(x(3)),cos(x(3)),
x(2);0 0 1]
Mch=Rch*Mch;
Mavd=Rch*[cos(x(5)),-sin(x(5)) 3;sin(x(5)),cos(x(5)) 3
;0 0 1]*Mav;
Mavg=Rch*[cos(x(5)),-sin(x(5)) 3;sin(x(5)),cos(x(5))
-3;0 0 1]*Mav;
plot(Mch(1, :),Mch(2, :),'blue');
plot(Mavd(1, :),Mavd(2, :),'black');
plot(Mavg(1, :),Mavg(2, :),'black');
```

Solution to Exercise 2.5 (simulation of a pendulum)

The state equations of the pendulum are:

$$\begin{pmatrix} \dot{x}_1 \\ \dot{x}_2 \end{pmatrix} = \begin{pmatrix} x_2 \\ \frac{-g\sin x_1}{\ell} \end{pmatrix}$$

The MATLAB program allowing us to simulate the pendulum during 1 s is given below. It computes an approximation of the vector x corresponding to the state of the pendulum for $t = 1s$. The angle corresponds to x_1.

```
L=1;g=9.81;dt=0.01; % initialization
x= [1;0]; % initial condition
for t=0:dt:1,
x=x+ [x(2); -(g/L)*sin(x(1))]*dt;
end;
```

This program is also given in the file pendule_main.m. It calls the evolution function pendule_f.m and the drawing function pendule_draw.m.

Solution to Exercise 2.6 (Van der Pol system)

1) The state equations are:

$$\begin{cases} \dot{x}_1 = x_2 \\ \dot{x}_2 = -\left(x_1^2 - 1\right) x_2 - x_1 \end{cases}$$

2) The point of equilibrium satisfies:

$$\begin{cases} 0 = x_2 \\ 0 = \left(x_1^2 - 1\right) x_2 + x_1 \end{cases}$$

i.e. $x_1 = x_2 = 0$. By linearizing around $\bar{\mathbf{x}} = (0\ 0)^\mathrm{T}$, we obtain:

$$\begin{cases} \dot{x}_1 = x_2 \\ \dot{x}_2 = x_2 - x_1 \end{cases}$$

Let:

$$\dot{\mathbf{x}} = \begin{pmatrix} 0 & 1 \\ -1 & 1 \end{pmatrix} \mathbf{x}$$

The characteristic polynomial is $s^2 - s + 1$. The eigenvalues of the evolution matrix are therefore $\frac{1}{2} \pm \frac{\sqrt{3}}{2}i$. The system is unstable.

3) The path is drawn in Figure 2.15.

We can observe a stable limit cycle. The output $y(t)$ is given in Figure 2.16.

The Van der Pol system is an oscillator. Whatever the initial conditions are, it quickly starts oscillating with constant frequency and amplitude.

4) No, it is not possible to create a loop since the system is deterministic. For each x, there is only one corresponding $\dot{x}$.

5) The system is structurally stable, in other words every infinitely small perturbation of the vector field leaves its stability properties unchanged. This property is common for robots, but not for conservative mechanical systems (such as, the frictionless pendulum: the smallest friction alters its stability properties). The integration schemes will all, therefore, probably give an equivalent behavior.

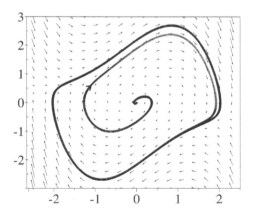

Figure 2.15. *Path for the Van der Pol system obtained by Euler's method*

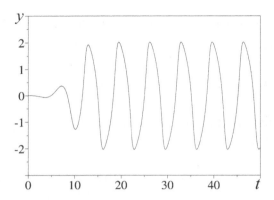

Figure 2.16. *Output $y(t)$ obtained for the Van der Pol oscillator*

Solution to Exercise 2.7 (simulation of a car)

The complete program of the simulation of the car is given below.

```
x= [0;0;0;7;0];u= [0;0.2];dt=0.01;
for t=0:dt:10,
xpoint=
[x(4)*cos(x(5))*cos(x(3));x(4)*cos(x(5))*sin(x(3));
x(4)*sin(x(5))/3; u(1);u(2)];
x=x+xpoint*dt;
end;
```

Solution to Exercise 2.8 (integration using Taylor's method)

We have:

$$\begin{pmatrix} \ddot{x}_1 \\ \ddot{x}_2 \\ \ddot{x}_3 \\ \ddot{x}_4 \\ \ddot{x}_5 \end{pmatrix} = \begin{pmatrix} \dot{x}_4 \cos x_3 - x_4 \dot{x}_3 \sin x_3 \\ \dot{x}_4 \sin x_3 + x_4 \dot{x}_3 \cos x_3 \\ \dot{x}_5 \\ \dot{u}_1 \\ \dot{u}_2 \end{pmatrix} = \begin{pmatrix} u_1 \cos x_3 - x_4 x_5 \sin x_3 \\ u_1 \sin x_3 + x_4 x_5 \cos x_3 \\ u_2 \\ \dot{u}_1 \\ \dot{u}_2 \end{pmatrix}$$

Thus, the second-order Taylor integration scheme is written as:

$$\underbrace{\begin{pmatrix} \hat{x}_1(t+dt) \\ \hat{x}_2(t+dt) \\ \hat{x}_3(t+dt) \\ \hat{x}_4(t+dt) \\ \hat{x}_5(t+dt) \end{pmatrix}}_{\hat{\mathbf{x}}(t+dt)} = \underbrace{\begin{pmatrix} \hat{x}_1(t) \\ \hat{x}_2(t) \\ \hat{x}_3(t) \\ \hat{x}_4(t) \\ \hat{x}_5(t) \end{pmatrix}}_{\hat{\mathbf{x}}(t)} + dt \cdot \underbrace{\begin{pmatrix} \hat{x}_4(t).\cos \hat{x}_3(t) \\ \hat{x}_4(t).\sin \hat{x}_3(t) \\ \hat{x}_5(t) \\ u_1(t) \\ u_2(t) \end{pmatrix}}_{\mathbf{f}(\hat{\mathbf{x}}(t),\mathbf{u}(t))}$$

$$+dt^2 \cdot \underbrace{\begin{pmatrix} u_1(t).\cos \hat{x}_3(t) - \hat{x}_4(t).\hat{x}_5(t).\sin \hat{x}_3(t) \\ u_1(t).\sin \hat{x}_3(t) + \hat{x}_4(t).\hat{x}_5(t).\cos \hat{x}_3(t) \\ u_2(t) \\ \dot{u}_1(t) \\ \dot{u}_2(t) \end{pmatrix}}_{\frac{\partial \mathbf{f}}{\partial \mathbf{x}}(\hat{\mathbf{x}}(t),\mathbf{u}(t)).\mathbf{f}(\hat{\mathbf{x}}(t),\mathbf{u}(t)) + \frac{\partial \mathbf{f}}{\partial \mathbf{u}}(\hat{\mathbf{x}}(t),\mathbf{u}(t)).\dot{\mathbf{u}}(t)}$$

In order to be able to apply this integration scheme, we must take:

$$\mathbf{u}(t) = \begin{pmatrix} \cos t \\ \sin t \end{pmatrix} \text{ et } \dot{\mathbf{u}}(t) = \begin{pmatrix} -\sin t \\ \cos t \end{pmatrix}$$

Solution to Exercise 2.9 (radius of a tricycle wheel)

This transformation matrix is given by:

$$\mathbf{R} := \underbrace{\begin{pmatrix} 1 & 0 & 0 & x \\ 0 & 1 & 0 & y \\ 0 & 0 & 1 & 0 \\ 0 & 0 & 0 & 1 \end{pmatrix}}_{\text{translation of the body}} \cdot \underbrace{\exp \begin{pmatrix} 0 & -\theta & 0 & 0 \\ \theta & 0 & 0 & 0 \\ 0 & 0 & 0 & 0 \\ 0 & 0 & 0 & 1 \end{pmatrix}}_{\text{rotation of the body}} \cdot \underbrace{\begin{pmatrix} 1 & 0 & 0 & L \\ 0 & 1 & 0 & 0 \\ 0 & 0 & 1 & r \\ 0 & 0 & 0 & 1 \end{pmatrix}}_{\text{positioning of the front wheel}}$$

$$\cdot \underbrace{\exp \begin{pmatrix} 0 & -\delta & 0 & 0 \\ \delta & 0 & 0 & 0 \\ 0 & 0 & 0 & 0 \\ 0 & 0 & 0 & 1 \end{pmatrix}}_{\text{rotation of } \delta \text{ with respect to } Oz} \cdot \underbrace{\exp \begin{pmatrix} 0 & 0 & \alpha & 0 \\ 0 & 0 & 0 & 0 \\ -\alpha & 0 & 0 & 0 \\ 0 & 0 & 0 & 1 \end{pmatrix}}_{\text{rotation of } \alpha \text{ with respect to } Oy}$$

Solution to Exercise 2.10 (three-dimensional simulation of a tricycle in MATLAB)

1) A matrix in homogeneous coordinates is composed of 4 rows and as much columns as there are points in the pattern. The last row only contains 1s. For instance for the unit cube, we have:

$$\mathbf{M} = \begin{pmatrix} 0 & 1 & 1 & 0 & 0 & 1 & 1 & 0 & 0 & 0 & 1 & 1 & 1 & 1 \\ 0 & 0 & 0 & 0 & 1 & 1 & 1 & 1 & 1 & 0 & 0 & 1 & 1 & 0 \\ 0 & 0 & 1 & 1 & 0 & 0 & 0 & 1 & 1 & 0 & 1 & 1 & 1 & 0 & 0 \\ 1 & 1 & 1 & 1 & 1 & 1 & 1 & 1 & 1 & 1 & 1 & 1 & 1 & 1 \end{pmatrix}$$

Rodrigues' formula tells us that the expression of the rotation matrix $\mathbf{R}$ of angle $\|\omega\|$ around ω is given by:

$$\mathbf{R} = \exp \begin{pmatrix} 0 & -\omega_z & \omega_y \\ \omega_z & 0 & -\omega_x \\ -\omega_y & \omega_x & 0 \end{pmatrix}$$

Thus, in order to rotate the pattern M then translate it by (t_x, t_y, t_y), we will perform the following operation:

$$\mathbf{M} := \underbrace{\begin{pmatrix} 1 & 0 & 0 & t_x \\ 0 & 1 & 0 & t_y \\ 0 & 0 & 1 & t_z \\ 0 & 0 & 0 & 1 \end{pmatrix}}_{\text{translation}} \cdot \underbrace{\begin{pmatrix} \exp \begin{pmatrix} 0 & -\omega_z & \omega_y \\ \omega_z & 0 & -\omega_x \\ -\omega_y & \omega_x & 0 \end{pmatrix} & \begin{matrix} 0 \\ 0 \\ 0 \end{matrix} \\ \begin{matrix} 0 & 0 & 0 \end{matrix} & 1 \end{pmatrix}}_{\text{rotation}} \cdot \mathbf{M}$$

2) The drawing function is given below.

```
function tricycle3D_draw(x,L,e,t)
M = [-1 2 1 0 -1 -1 2 1 0 -1 0 0 1 1 2 L 2 0 0 0; ...
-1 -1 -0.5 -0.5 -1 1 1 0.5 0.5 1 0.5 -0.5 -0.5 0.5 1 0 -1
-1 -e e; ...
0 0 1 1 0 0 0 1 1 0 1 1 1 1 0 0 0 0 0 0; ...
1 1 1 1 1 1 1 1 1 1 1 1 1 1 1 1 1 1 1 1];
```

```
wheel = [ ];
for a=0:pi/8:2*pi; % pattern of the wheel
radius = [cos(a); 0; sin(a); 1];
wheel = [wheel,radius, [0;0;0;1],radius];
end
Rcam = Rotate([0.8;0;0]); % position the camera
Rchassis = Rcam*Translate([x(1);x(2);0])*Rotate
     ([0;0;x(3)]); M1 = Rchassis*M;
wheel1 = Rchassis*Translate([0;-2;0])*Rotate
     ([0;x(6);0])*wheel;
wheel2 = Rchassis*Translate([0; 2;0])*Rotate
     ([0;x(7);0])*wheel;
wheel3 = Rchassis*Translate([3; 0;0])*Rotate([0;0;x(5)])
     *Rotate([0;x(8);0])*wheel;
axis([-7,7,-5,5]); % axes
plot(M1(1, :),M1(3, :),'black'); % chassis
plot(roue1(1, :),roue1(3, :),'blue'); % wheel 1
plot(roue2(1, :),roue2(3, :),'blue'); % wheel 2
plot(roue3(1, :),roue3(3, :),'green'); % wheel 3
```

3) Following the speed composition rule, we have, for the first wheel:

$$\mathbf{v}_{R_1} = \mathbf{v}_A + \overrightarrow{R_1 A} \wedge \vec{\omega}$$

or, by expressing this relation in the frame of the tricycle (see figure of the problem statement):

$$\begin{pmatrix} r\dot{\alpha}_1 \\ 0 \\ 0 \end{pmatrix} = \begin{pmatrix} v\cos\delta \\ v\sin\delta \\ 0 \end{pmatrix} + \begin{pmatrix} L \\ e \\ 0 \end{pmatrix} \wedge \begin{pmatrix} 0 \\ 0 \\ \dot{\theta} \end{pmatrix}$$

or:
$$r\dot{\alpha}_1 = v\cos\delta - e\dot{\theta} = v\cos\delta + e\frac{v\sin\delta}{L}$$

Thus:
$$\dot{\alpha}_1 = \frac{v}{r}\left(\cos\delta + \frac{e\sin\delta}{L}\right)$$

For the second wheel, the same reasoning applies. We obtain:
$$\dot{\alpha}_2 = \frac{v}{r}\left(\cos\delta - \frac{e\sin\delta}{L}\right)$$

For the third wheel, we have:
$$\dot{\alpha}_3 = \frac{v}{r}$$

4) We obtain:
$$\begin{pmatrix} \dot{x} \\ \dot{y} \\ \dot{\theta} \\ \dot{v} \\ \dot{\delta} \\ \dot{\alpha}_1 \\ \dot{\alpha}_2 \\ \dot{\alpha}_3 \end{pmatrix} = \begin{pmatrix} v\cos\delta\cos\theta \\ v\cos\delta\sin\theta \\ \frac{v\sin\delta}{L} \\ u_1 \\ u_2 \\ \frac{v}{r}\left(\cos\delta + \frac{e\sin\delta}{L}\right) \\ \frac{v}{r}\left(\cos\delta - \frac{ev\sin\delta}{L}\right) \\ \frac{v}{r} \end{pmatrix}$$

The simulation program is given below.

```
L = 3; % distance between the axles
e = 2; % distance between the wheel and the car axis
x = [-1;-1;0.8;1;0.3;0;0;0];
% x = (x,y,theta,v,delta,a1,a2,a3)
```

```
% ai is the angle of wheel i
dt = 0.01;
for t = 0:dt:5;
tricycle3D_draw(x,L,e,t);
u = [0.1;0.1];
x = x+tricycle3D_f(x,u,L,e)*dt;
end;
```

This program is available in the file `tricycle3D_main.m`.

Solution to Exercise 2.11 (manipulator arms)

Table 2.1 represents the series of transformations to apply to the pattern in order to represent each arm. As indicated in this table, the second arm has to be subjected to all the transformations applied to the first arm and the third arm to those applied to the second arm.

Arm 1	Arm 2	Arm 3
$\mathrm{Diag}(3, 0.3, 0.3)$	$\mathrm{Diag}(2, 0.3, 0.3)$	$\mathrm{Diag}(1, 0.3, 0.3)$
$\mathrm{Rot}_z(\alpha_1)$	$\mathrm{Rot}_z(\alpha_2)$	$\mathrm{Rot}_x(\alpha_3)$
$\mathrm{Rot}_x(0.5)$	$\mathrm{Trans}(3, 0, 0)$	$\mathrm{Trans}(2, 0, 0)$
	$\mathrm{Rot}_z(\alpha_1)$	$\mathrm{Rot}_z(\alpha_2)$
	$\mathrm{Rot}_x(0.5)$	$\mathrm{Trans}(3, 0, 0)$
		$\mathrm{Rot}_z(\alpha_1)$
		$\mathrm{Rot}_x(0.5)$

Table 2.1. *Transformation for the three arms of the robot*

The MATLAB program (see file `bras3D.m`) below simulates the movement of these arms.

```
M0 = [0 1 1 0 0 0 1 1 0 0 0 1 1 1 1; ...
0 0 0 0 0 1 1 1 1 1 0 0 1 1 0; ...
0 0 1 1 0 0 0 1 1 0 1 1 1 0 0; ...
1 1 1 1 1 1 1 1 1 1 1 1 1 1 1] ;
M0=Translate([0;-0.5;-0.5])*M0;
Rcam=Rotate([0.5;0;0]); % camera
```

```
x = [0;0;0]; % initial state
dt = 0.005;
w = [2;-4;5]; % angular speeds
for t=0:dt:3,
M1=diag([3,0.3,0.3,1])*M0; R1=Rcam*Rotate([0;0;
x(1)]);
M2=diag([2,0.3,0.3,1])*M0; R2=Translate([3;0;
0])*Rotate([0;0;x(2)]);
M3=diag([1,0.3,0.3,1])*M0; R3=Translate([2;0;
0])*Rotate([x(3);0;0]);
M1=R1*M1; M2=R1*R2*M2; M3=R1*R2*R3*M3;
plot(M1(1, :), M1(2, :), '-blue.');
plot(M2(1, :), M2(2, :), '-green.');
plot(M3(1, :), M3(2, :), '-black.');
x = x + dt*w;
end;
```

Solution to Exercise 2.12 (simulation and control of a skating vehicle in MATLAB*)*

1) The angular speed of the sledge is given by:

$$\dot{\theta} = \frac{v_1 \sin \beta}{L_1}$$

where v_1 is the speed of the front skate and L_1 is the distance between the front skate and the center of the axle of the front sledge. If v corresponds to the speed of the middle of the front sledge axle, we have $v = v_1 \cos \beta$. These two relations give us:

$$\dot{\theta} = \frac{v \tan \beta}{L_1} = \frac{vu_1}{L_1}$$

From the point of view of the rear sledge, everything happens as if there was a virtual wheel in the middle of the front sledge axle that moves together with the latter. Thus, by taking the formula above, the angular speed of the rear sledge is $\dot{\theta}+\dot{\delta}=-\frac{v\sin\delta}{L_2}$ where L_2 is the distance between the centers of the axles. And therefore:

$$\dot{\delta}=-\frac{v\sin\delta}{L_2}-\dot{\theta}=-\frac{v\sin\delta}{L_2}-\frac{vu_1}{L_1}$$

Following the theorem of kinetic energy, the temporal derivative of the kinetic energy is equal to the sum of the powers brought in to the system, i.e.:

$$\frac{d}{dt}\left(\frac{1}{2}mv^2\right)=\underbrace{u_2.\dot{\delta}}_{\text{engine power}}-\underbrace{(\alpha v).v}_{\text{dissipated power}}$$

where α is the coefficient of viscous friction. For reasons of simplification, we have just assumed that the friction force is αv, which assumes that only the front sledge brakes. We then have:

$$mv\dot{v}=u_2.\dot{\delta}-\alpha v^2=u_2.\left(-\frac{v\sin\delta}{L_2}-\frac{vu_1}{L_1}\right)-\alpha v^2$$

or:

$$m\dot{v}=u_2.\left(-\frac{\sin\delta}{L_2}-\frac{u_1}{L_1}\right)-\alpha v$$

The system can be described by the following state equations:

$$\begin{cases}\dot{x}=v\cos\theta\\ \dot{y}=v\sin\theta\\ \dot{\theta}=vu_1\\ \dot{v}=-\left(u_1+\sin\delta\right)u_2-v\\ \dot{\delta}=-v\left(u_1+\sin\delta\right)\end{cases}$$

or, to simplify, the coefficients (mass m, coefficient of viscous friction α, interaxle distances L_1, L_2, etc.) have been chosen unitary.

2) First of all, we create in the file `snake_f.m` the following evolution function:

```
function v=snake_f(x,u)
theta=x(3); v=x(4); delta=x(5);
v= [v*cos(theta); v*sin(theta); v*u(1); ...
-u(2)*(u(1)+sin(delta))-v; -v*(u(1)+sin(delta))];
```

Then, we program Euler's method in the file `snake_main.m` as follows:

```
x= [0;0;0;3;0.5]; dt=0.05;
for t=0:dt:2,
u= [1;0];
x=x+snake_f(x,u)*dt;
snake_draw(x,u);
end;
```

3) We choose u_2 in order for the propulsion torque to be a driving force, in other words $\dot{\delta} u_2 \geq 0$. Indeed, $\dot{\delta} u_2$ corresponds to the power given to the robot that is transformed into kinetic energy. If u_2 is bounded by the interval $[-p_4, p_4]$, we choose bang-bang-type control for u_2 of the form:

$$u_2 = p_4 \text{sign}\left(\dot{\delta}\right)$$

which means to exert a maximum propulsion. The chosen control is therefore:

$$\mathbf{u} = \begin{pmatrix} p_1 \cos(p_2 t) + p_3 \\ p_4 \text{sign}(-v(u_1 + \sin \delta)) \end{pmatrix}$$

The parameters of the control remain to be determined. The bias parameter p_3 is allowed to vary. The engine torque power gives us p_4. The parameter p_1 is directly linked to the amplitude of the oscillation created during the movement. Finally, the parameter p_2 gives the frequency of the oscillations. The simulations can help us to correctly fixate parameters p_1 and p_2. Figure 2.13 illustrates two simulations where the robot starts with a quasi-nil speed. In the upper simulation, the bias p_3 is nil. In the lower simulation, we have $p_3 > 0$. The main program (see file snake_main.m) is given below:

```
x= [0;0;2;0.1;0];   % x,y,theta,v,delta
dt=0.05;p1=0.5;p2=3;p3=0;p4=5;thetabar=pi/6;
for t=0:dt:10,
u1=p1*cos(p2*t)+p3;
u= [u1;p4*sign(-x(4)*(u1+sin(x(5))))];
x=x+snake_f(x,u)*dt;
snake_draw(x,u);
end;
```

4) We control the bias p_3 by a sawtooth-type proportional control of the form:

$$p_3 = \operatorname{atan}\left(\tan\frac{\bar{\theta} - \theta}{2}\right)$$

in order to avoid jumps of 2π. In MATLAB: p3=atan(tan((thetabar-x(3))/2));

3

Linear Systems

The study of linear systems [BOU 06] is fundamental for the proper understanding of the concepts of stability and the design of linear controllers. Let us recall that linear systems are of the form:

$$\begin{cases} \dot{\mathbf{x}}(t) = \mathbf{A}\mathbf{x}(t) + \mathbf{B}\mathbf{u}(t) \\ \mathbf{y}(t) = \mathbf{C}\mathbf{x}(t) + \mathbf{D}\mathbf{u}(t) \end{cases}$$

for continuous-time systems and:

$$\begin{cases} \mathbf{x}(k+1) = \mathbf{A}\mathbf{x}(k) + \mathbf{B}\mathbf{u}(k) \\ \mathbf{y}(k) = \mathbf{C}\mathbf{x}(k) + \mathbf{D}\mathbf{u}(k) \end{cases}$$

for discrete-time systems.

3.1. Stability

A linear system is *stable* (also called *asymptotically stable* in the literature) if, after a sufficiently long period of time, the state no longer depends on the initial conditions, no matter what they are. This means (see Exercises 3.1 and 3.2) that:

$\lim_{t \to \infty} e^{\mathbf{A}t} = \mathbf{0}_n$ if the system is continuous-time

$\lim_{k \to \infty} \mathbf{A}^k = \mathbf{0}_n$ if the system is discrete-time

In this expression, we can see the concept of *matrix exponential*. The exponential of a square matrix M of dimension n can be defined through its integer series development:

$$e^{\mathbf{M}} = \mathbf{I}_n + \mathbf{M} + \frac{1}{2!}\mathbf{M}^2 + \frac{1}{3!}\mathbf{M}^3 + \cdots = \sum_{i=0}^{\infty} \frac{1}{i!}\mathbf{M}^i$$

where $\mathbf{I}_n$ is the identity matrix of dimension n. It is clear that $e^{\mathbf{M}}$ is of the same dimension as M. Here are some of the important properties concerning the exponentials of matrices. If $\mathbf{0}_n$ is the zero matrix of $n \times n$ and if M and N are two matrices $n \times n$, then:

$$e^{\mathbf{0}_n} = \mathbf{I}_n$$
$$e^{\mathbf{M}}.e^{\mathbf{N}} = e^{\mathbf{M}+\mathbf{N}} \text{ (if the matrices commute)}$$
$$\frac{d}{dt}\left(e^{\mathbf{M}t}\right) = \mathbf{M}e^{\mathbf{M}t}$$

CRITERION OF STABILITY.– There is a criterion of stability that only depends on the matrix A. A continuous-time linear system is stable if and only if all the eigenvalues of its evolution matrix A have strictly negative real parts. A discrete-time linear system is stable if and only if all the eigenvalues of A are strictly inside the unit circle. This corresponds to the criterion of stability as proven in Exercise 3.3.

Whether it is for continuous-time or discrete-time linear systems, the position of the eigenvalues of A is of paramount importance in the study of stability of a linear system. The *characteristic polynomial* of a linear system is defined as the characteristic polynomial of the matrix A that is given by the formula:

$$P(s) = \det\left(s\mathbf{I}_n - \mathbf{A}\right)$$

Its roots are the eigenvalues of A. Indeed, if s is a root of $P(s)$, then $\det\left(s\mathbf{I}_n - \mathbf{A}\right) = 0$, in other words there is a vector v non-nil such that $(s\mathbf{I}_n - \mathbf{A})\mathbf{v} = \mathbf{0}$. This means that $s\mathbf{v} - \mathbf{A}\mathbf{v} = \mathbf{0}$ or $\mathbf{A}\mathbf{v} = s\mathbf{v}$. Therefore s is an eigenvalue of $\mathbf{A}$. A corollary of the stability criterion is thus the following.

COROLLARY.− A continuous-time linear system is stable if and only if all the roots of its characteristic polynomial have negative real parts. A discrete-time linear system is stable if and only if all the roots of its characteristic polynomial are inside the unit circle.

3.2. Laplace transform

The *Laplace transform* is a very useful tool for the control engineer in the manipulation of systems described by differential equations. This approach is particularly utilized in the context of monovariate systems (i.e. systems with a single input and output) and may be regarded as a competitor to the state-representation approach considered in this book.

3.2.1. *Laplace variable*

The space of differential operators in $\frac{d}{dt}$ is a ring and has favorable properties such as associativity. For example:

$$\frac{d^4}{dt^4}\left(\frac{d^3}{dt^3} + \frac{d}{dt}\right) = \frac{d^4}{dt^4}\frac{d^3}{dt^3} + \frac{d^4}{dt^4}\frac{d}{dt} = \frac{d^7}{dt^7} + \frac{d^5}{dt^5}$$

This ring is commutative. For example:

$$\frac{d^4}{dt^4}\left(\frac{d^3}{dt^3} + \frac{d}{dt}\right) = \left(\frac{d^3}{dt^3} + \frac{d}{dt}\right)\frac{d^4}{dt^4}$$

We may associate with the operator $\frac{d}{dt}$ the symbol s called *Laplace variable*. Thus the operator $\frac{d^4}{dt^4}\left(\frac{d^3}{dt^3} + \frac{d}{dt}\right)$ will be represented by the polynomial $s^4\left(s^3 + s\right)$.

3.2.2. *Transfer function*

Let us consider a linear system with input u and output y described by a differential relation such as:

$$y(t) = H\left(\frac{d}{dt}\right).u(t)$$

The function $H(s)$ is called the *transfer function* of the system. Let us take, for instance, the system described by the differential equation:

$$\ddot{y} + 2\dot{y} + 3y = 4\dot{u} - 5u$$

We have:

$$y(t) = \left(\frac{4\frac{d}{dt} - 5}{\frac{d^2}{dt^2} + 2\frac{d}{dt} + 3}\right).u(t)$$

Its transfer function is therefore:

$$H(s) = \frac{4s - 5}{s^2 + 2s + 3}$$

If the transfer function $H(s)$ of a system is a rational function, its denominator $P(s)$ is called the *characteristic polynomial*.

3.2.3. *Laplace transform*

We call *Laplace transform* $\hat{y}(s)$ of the signal $y(t)$ the transfer function of the system that generates $y(t)$ from the Dirac delta function $\delta(t)$. We will denote it by $\hat{y}(s) = \mathcal{L}(y(t))$. We also say that $y(t)$ is the *impulse response* of the system. Table 3.1 shows several systems together with their transfer function and impulse response.

In this table, $E(t)$ is the unit step which is equal to 1 if $t \geq 0$ and 0 otherwise. Thus, the Laplace transforms of

$\delta(t), E(t), \dot{\delta}(t), \delta(t-\tau)$ are respectively $1, \frac{1}{s}, s, e^{-\tau s}$. But we can go further, as is shown in Table 3.2.

System	Equation	Transfer function	Impulse response
Identity	$y(t) = u(t)$	1	$\delta(t)$
Integrator	$\dot{y}(t) = u(t)$	$\frac{1}{s}$	$E(t)$ (step)
Differentiator	$y(t) = \dot{u}(t)$	s	$\dot{\delta}(t)$
Delay	$y(t) = u(t-\tau)$	$e^{-\tau s}$	$\delta(t-\tau)$

Table 3.1. *Transfer function and impulse response of some elementary systems*

Equation	Transfer function	Impulse response
$y(t) = \alpha_1 u(t-\tau_1) + \alpha_2 u(t-\tau_2)$	$\alpha_1 e^{-\tau_1 s} + \alpha_2 e^{-\tau_2 s}$	$\alpha_1 \delta(t-\tau_1) + \alpha_2 \delta(t-\tau_2)$
$y(t) = \sum_{i=0}^{\infty} \alpha_i u(t-\tau_i)$	$\sum_{i=0}^{\infty} \alpha_i e^{-\tau_i s}$	$\sum_{i=0}^{\infty} \alpha_i \delta(t-\tau_i)$
$y(t) = \int_{-\infty}^{\infty} f(\tau) u(t-\tau) d\tau$	$\int_{-\infty}^{\infty} f(\tau) e^{-\tau s} d\tau$	$\int_{-\infty}^{\infty} f(\tau) \delta(t-\tau) d\tau$

Table 3.2. *Transfer function and impulse response of composed systems*

The operation $y(t) = \int_{-\infty}^{\infty} f(\tau) u(t-\tau)$ is called *convolution*. We may notice that the impulse response of the system described by the last row of the table is:

$$y(t) = \int_{-\infty}^{\infty} f(\tau) u(t-\tau) d\tau \bigg|_{u(t)=\delta(t)}$$
$$= \int_{-\infty}^{\infty} f(\tau) \delta(t-\tau) d\tau = f(t)$$

and therefore the Laplace transform of a function $f(t)$ is:

$$\hat{f}(s) = \int_{-\infty}^{\infty} f(\tau) e^{-\tau s} d\tau$$

Let us note that the relation:

$$f(t) = \int_{-\infty}^{\infty} f(\tau)\delta(t-\tau)d\tau$$

illustrates the fact that $f(t)$ can be estimated by the sum of an infinity of infinitely approached Dirac distributions.

3.2.4. *Input–output relation*

Let us consider a system with input u, output y and transfer function $H(s)$, as in Figure 3.1.

Figure 3.1. *Laplace transform and transfer function*

We have:

$$y(t) = H(\frac{d}{dt}).u(t) = \underbrace{H(\frac{d}{dt}).\hat{u}\left(\frac{d}{dt}\right).\delta(t)}_{\hat{y}(\frac{d}{dt})}$$

Therefore, the Laplace transform of $y(t)$ is:

$$\hat{y}(s) = H(s).\hat{u}(s)$$

3.3. Relationship between state and transfer representations

Let us consider the system described by its state equations:

$$\begin{cases} \dot{\mathbf{x}} = \mathbf{A}\mathbf{x} + \mathbf{B}\mathbf{u} \\ \mathbf{y} = \mathbf{C}\mathbf{x} + \mathbf{D}\mathbf{u} \end{cases}$$

The Laplace transform of the state representation is given by:

$$\begin{cases} s\hat{x} = A\hat{x} + B\hat{u} \\ \hat{y} = C\hat{x} + D\hat{u} \end{cases}$$

The first equation can be re-written as $s\hat{x} - A\hat{x} = B\hat{u}$, i.e. $sI\hat{x} - A\hat{x} = B\hat{u}$, where I is the identity matrix. From this, by factoring, we get $(sI - A)\hat{x} = B\hat{u}$ (we must be careful, a notation such as $s\hat{x} - A\hat{x} = (s - A)\hat{x}$ is not permitted since s is a scalar whereas A is a matrix). Therefore:

$$\begin{cases} \hat{x} = (sI - A)^{-1} B\hat{u} \\ \hat{y} = C\hat{x} + D\hat{u} \end{cases}$$

and thus:

$$\hat{y} = \left(C(sI - A)^{-1} B + D \right) \hat{u}$$

The matrix:

$$G(s) = C(sI - A)^{-1} B + D$$

is called *transfer matrix*. It is a matrix of *transfer functions* (in other words of rational functions in s) in which every denominator is a divisor of the characteristic polynomial $P_A(s)$ of A. By multiplying each side by $P_A(s)$ and replacing s by $\frac{d}{dt}$, we obtain a system of input–output differential equations. The state x will no longer appear there.

Reciprocally, in the case of *monovariate* linear systems (in other words with a single input and a single output), we can obtain a state representation from an expression of the transfer function, as it will be illustrated in Exercises 3.16, 3.17 and 3.18.

3.4. Exercises

EXERCISE 3.1.– Solution of a continuous-time linear state equation

Show that the continuous-time linear system:

$$\dot{\mathbf{x}}(t) = \mathbf{A}\mathbf{x}(t) + \mathbf{B}\mathbf{u}(t)$$

has the solution:

$$\mathbf{x}(t) = e^{\mathbf{A}t}\mathbf{x}(0) + \int_0^t e^{\mathbf{A}(t-\tau)}\mathbf{B}\mathbf{u}(\tau)d\tau$$

The function $e^{\mathbf{A}t}\mathbf{x}(0)$ is called *homogeneous, free* or *transient solution*. The function $\int_0^t e^{\mathbf{A}(t-\tau)}\mathbf{B}\mathbf{u}(\tau)d\tau$ is called *forced solution*.

EXERCISE 3.2.– Solution of a discrete-time linear state equation

Show that the discrete-time linear system:

$$\mathbf{x}(k+1) = \mathbf{A}\mathbf{x}(k) + \mathbf{B}\mathbf{u}(k)$$

has the solution:

$$\mathbf{x}(k) = \mathbf{A}^k\mathbf{x}(0) + \sum_{\ell=0}^{k-1} \mathbf{A}^{k-\ell-1}\mathbf{B}\mathbf{u}(\ell)$$

The function $\mathbf{A}^k\mathbf{x}(0)$ is the *homogeneous solution*, and the function $\sum_{\ell=0}^{k} \mathbf{A}^{k-\ell}\mathbf{B}\mathbf{u}(\ell)$ is the *forced solution*.

EXERCISE 3.3.– Criterion of stability

The proof of the stability criterion for linear systems is based on the theorem of correspondence of the eigenvalues that is articulated as follows. If f is a polynomial (or more generally an integer series) and if $\mathbf{A}$ is an $\mathbb{R}^{n \times n}$ matrix then the eigenvectors of $\mathbf{A}$ are also eigenvectors of $f(\mathbf{A})$. Moreover if the eigenvalues of $\mathbf{A}$ are $\{\lambda_1, \ldots, \lambda_n\}$ then those of $f(\mathbf{A})$ are $\{f(\lambda_1), \ldots, f(\lambda_n)\}$.

1) Prove the theorem of correspondence of the eigenvalues in the case where f is a polynomial.

2) Let $\mathrm{spec}(\mathbf{A}) = \{\lambda_1, \ldots, \lambda_n\}$ be the spectrum of $\mathbf{A}$, i.e. its eigenvalues. By using the theorem of correspondence of the eigenvalues, calculate $\mathrm{spec}(\mathbf{I} + \mathbf{A})$, $\mathrm{spec}(\mathbf{A}^k)$, $\mathrm{spec}(e^{\mathbf{A}t})$ and $\mathrm{spec}(f(\mathbf{A}))$. In these expressions, $\mathbf{I}$ denotes the identity matrix, k is an integer and f is the characteristic polynomial of $\mathbf{A}$.

3) Show that if a continuous-time linear system is stable, then all the eigenvalues of its evolution matrix $\mathbf{A}$ have strictly negative real parts (in fact, the condition is necessary and sufficient, but we will limit the proof to the implication).

4) Show that if a discrete-time linear system is stable, then all the eigenvalues of $\mathbf{A}$ are strictly within the unit circle. Again, even though we have the equivalence here, we will limit ourselves to the proof of the implication.

EXERCISE 3.4.– Laplace variable

Let us consider two systems in parallel as illustrated in Figure 3.2.

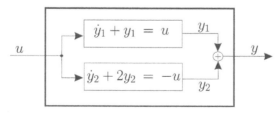

Figure 3.2. *Two linear systems interconnected in parallel*

1) Without employing the Laplace transform, and by only using elementary differential calculus, give a differential

equation that links u with y.

2) Using algebraic manipulations involving the $\frac{d}{dt}$ operator, obtain the same differential equation again.

3) Obtain the same result by using the Laplace variable s.

EXERCISE 3.5.– Transfer functions of elementary systems

Give the transfer function of the following systems with input u and output y:

1) a differentiator expressed by the differential equation $y = \dot{u}$;

2) an integrator that obeys the differential equation $\dot{y} = u$;

3) a delay of τ that is expressed by the input-output relation $y(t) = u(t - \tau)$.

EXERCISE 3.6.– Transfer function of composite systems

1) Let us consider two systems of transfer functions $H_1(s)$ and $H_2(s)$ placed in series as shown in Figure 3.3.

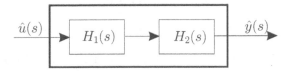

Figure 3.3. *Two systems in series*

Calculate, in function of $H_1(s)$ and $H_2(s)$, the transfer function of the composite system.

2) Let us place the two systems of transfer functions $H_1(s)$ and $H_2(s)$ in parallel as shown in Figure 3.4.

Give, in function of $H_1(s)$ and $H_2(s)$, the transfer function of the composite system.

3) Let us loop the system $H(s)$ as shown in Figure 3.5.

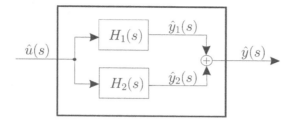

Figure 3.4. *Two systems in parallel*

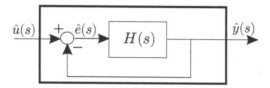

Figure 3.5. *Looped system*

Give, in function of $H(s)$, the transfer function of the looped system.

EXERCISE 3.7.– Transfer matrix

Let us consider the continuous-time linear system described by its state representation:

$$\begin{cases} \dot{\mathbf{x}}(t) = \begin{pmatrix} 1 & 3 \\ 2 & 0 \end{pmatrix} \mathbf{x}(t) + \begin{pmatrix} 1 \\ 1 \end{pmatrix} u(t) \\ \mathbf{y}(t) = \begin{pmatrix} 1 & 2 \\ 1 & 0 \end{pmatrix} \mathbf{x}(t) + \begin{pmatrix} 2 \\ 0 \end{pmatrix} u(t) \end{cases}$$

1) Calculate its transfer matrix.

2) Give a differential relation that links the input to the outputs.

EXERCISE 3.8.– Matrix block multiplication

The goal of this exercise is to recall the concept of block manipulation of matrices. This type of manipulation is widely used in linear control. We consider the following product of two block matrices:

$$\underbrace{\begin{pmatrix} C_{11} & C_{12} & C_{13} \\ C_{21} & C_{22} & C_{23} \end{pmatrix}}_{=C} = \underbrace{\begin{pmatrix} A_{11} & A_{12} & A_{13} \\ A_{21} & A_{22} & A_{23} \end{pmatrix}}_{=A} \cdot \underbrace{\begin{pmatrix} B_{11} & B_{12} & B_{13} \\ B_{21} & B_{22} & B_{23} \\ B_{31} & B_{32} & B_{33} \end{pmatrix}}_{=B}$$

As illustrated in Figure 3.6, we have:

$$C_{12} = A_{11} \cdot B_{12} + A_{12} \cdot B_{22} + A_{13} \cdot B_{32}$$

or, more generally:

$$C_{ij} = \sum_{k} A_{ik} \cdot B_{kj}$$

1) Write the conditions on the number of rows and columns for each sub-matrix so that the block product is possible.

2) The matrices C and A_{11} are square. The matrices A_{22} and B_{33} all have 3 rows of 2 columns. Find the dimensions of all the submatrices.

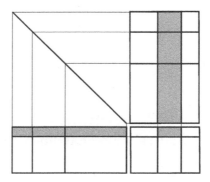

Figure 3.6. *Block multiplication of two matrices*

EXERCISE 3.9.– Change of basis

Let us consider the continuous-time linear system described by its state equations:

$$\begin{cases} \dot{\mathbf{x}} = \mathbf{A}\mathbf{x} + \mathbf{B}u \\ y = \mathbf{C}\mathbf{x} + \mathbf{D}u \end{cases}$$

Let us take the change of basis $\mathbf{v} = \mathbf{P}^{-1}\mathbf{x}$, where $\mathbf{P}$ is a transfer matrix (i.e. square and invertible).

1) What do the system equations become if $\mathbf{v}$ is the new state vector?

2) Let us consider the system described by the following state equations:

$$\begin{cases} \dot{\mathbf{x}} = \begin{pmatrix} 4 & -\frac{1}{2} & -\frac{1}{2} \\ 4 & 1 & -1 \\ 4 & -2 & 2 \end{pmatrix} \mathbf{x} + \begin{pmatrix} 1 \\ 2 \\ 4 \end{pmatrix} u \\ y = \begin{pmatrix} 2 & 1 & 1 \end{pmatrix} \mathbf{x} \end{cases}$$

We are trying to find a simpler representation, i.e. with more zeros and ones (in order for instance to limit the number of components that are necessary in the design of the circuitry). We propose to take the following change of basis:

$$\mathbf{v} = \begin{pmatrix} 1 & 1 & 1 \\ 2 & 1 & 2 \\ 2 & 1 & 0 \end{pmatrix}^{-1} \mathbf{x}$$

which brings us to a Jordan normal form. What does the new stat representation become?

3) Calculate the transfer function of this system. What is the characteristic polynomial of the system?

EXERCISE 3.10.– Change of basis toward a companion matrix

Let us consider a system with the input described by the evolution equation:

$$\dot{\mathbf{x}} = \mathbf{A}\mathbf{x} + \mathbf{b}u$$

Let us note that the control matrix traditionally denoted by B becomes, in our case of a single input, a vector b. Let us take as transfer matrix (assumed invertible):

$$\mathbf{P} = \left(\mathbf{b} \mid \mathbf{A}\mathbf{b} \mid \mathbf{A}^2\mathbf{b} \mid \ldots \mid \mathbf{A}^{n-1}\mathbf{b}\right)$$

The new state vector is therefore $\mathbf{v} = \mathbf{P}^{-1}\mathbf{x}$. Show that, in this new basis, the state equations are written as:

$$\dot{\mathbf{v}} = \begin{pmatrix} 0 & 0 & 0 & -a_0 \\ 1 & 0 & 0 & -a_1 \\ 0 & \vdots & 0 & \ldots \\ 0 & 0 & 1 & -a_{n-1} \end{pmatrix} \mathbf{v} + \begin{pmatrix} 1 \\ 0 \\ \vdots \\ 0 \end{pmatrix} u$$

where the a_i are the coefficients of the characteristic polynomial of the matrix **A**.

EXERCISE 3.11.– Pole-zero cancellation

Let us consider the system described by the state equation:

$$\begin{cases} \dot{x} = -x \\ y = x + u \end{cases}$$

Calculate the differential equation that links y to u, by the differential method, then by Laplace's method (without using pole-zero cancellation). What do you conclude?

EXERCISE 3.12.– State equations of a wiring system

Let us consider the system S described by the wiring system in Figure 3.7.

1) Give its state equations in matrix form.

2) Calculate the characteristic polynomial of the system. Is the system stable?

3) Calculate the transfer function of the system.

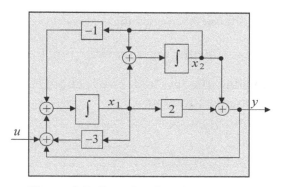

Figure 3.7. *Second-order wiring system*

EXERCISE 3.13.– Combination of systems

Let us consider the systems S_1 and S_2 given on top of Figure 3.8.

1) Give the state equations in matrix form of the system S_a obtained by placing systems S_1 and S_2 in series. Give the transfer function and characteristic polynomial of S_a.

2) Do the same with the system S_b obtained by placing systems S_1 and S_2 in parallel.

3) Do the same with the system S_c obtained by looping S_1 by S_2 as represented in Figure 3.8.

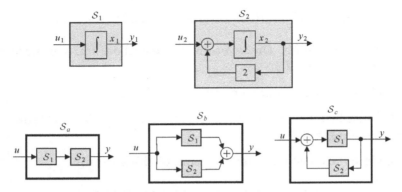

Figure 3.8. *Composition of systems*

EXERCISE 3.14.– Calculating a transfer function

Let us consider the system represented in Figure 3.9. Calculate its transfer function.

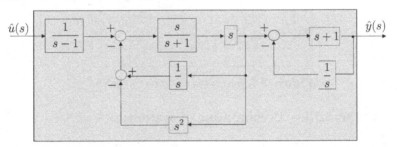

Figure 3.9. *Linear system for which the transfer function must be calculated*

EXERCISE 3.15.– Transfer matrix

Let us consider the continuous-time linear system described by its state representation:

$$\begin{cases} \dot{\mathbf{x}}(t) = \begin{pmatrix} 1 & 3 \\ 2 & 0 \end{pmatrix} \mathbf{x}(t) + \begin{pmatrix} 1 \\ 1 \end{pmatrix} u(t) \\ \mathbf{y}(t) = \begin{pmatrix} 1 & 2 \\ 1 & 0 \end{pmatrix} \mathbf{x}(t) + \begin{pmatrix} 2 \\ 0 \end{pmatrix} u(t) \end{cases}$$

1) Calculate its transfer function $G(s)$.

2) Deduce from this, a differential input–output relation for this system.

EXERCISE 3.16.– Canonical form of a control

Let us consider the order 3 linear system with a single input and a single output, described by the following differential equation:

$$\dddot{y} + a_2\ddot{y} + a_1\dot{y} + a_0 y = b_2\ddot{u} + b_1\dot{u} + b_0 u$$

1) Calculate its transfer function $G(s)$.

2) By noticing that this system can be obtained by placing the two transfer function systems in series:

$$G_1(s) = \frac{1}{s^3 + a_2 s^2 + a_1 s + a_0} \quad \text{and} \quad G_2(s) = b_2 s^2 + b_1 s + b_0$$

deduce a wiring system with only three integrators, some adders and amplifiers.

3) Give the state equations associated with this wiring.

EXERCISE 3.17.– Canonical observation form

Let us consider once more the system described by the differential equation:

$$\dddot{y} + a_2\ddot{y} + a_1\dot{y} + a_0 y = b_2\ddot{u} + b_1\dot{u} + b_0 u$$

1) Show that this differential equation can be written in integral form as:

$$y = \int\int\int (b_2\ddot{u} - a_2\ddot{y} + b_1\dot{u} - a_1\dot{y} + b_0 u - a_0 y)$$

2) Deduce from this a wiring system with only three integrators, some adders and amplifiers.

3) Give the state equations associated with this wiring.

4) Compare this with the results obtained in Exercise 3.16.

EXERCISE 3.18.– Modal form

A monovariate linear system is in *modal* form if it can be written as:

$$\begin{cases} \dot{\mathbf{x}} = \begin{pmatrix} \lambda_1 & 0 & \cdots & 0 \\ 0 & \lambda_2 & 0 & \cdots \\ \cdots & \cdots & \cdots & \cdots \\ 0 & \cdots & 0 & \lambda_n \end{pmatrix} \mathbf{x} + \begin{pmatrix} 1 \\ 1 \\ \vdots \\ 1 \end{pmatrix} u \\ y = \begin{pmatrix} c_1 & c_2 & \cdots & c_n \end{pmatrix} \mathbf{x} + d\, u \end{cases}$$

1) Draw a wiring system for this system using integrators, adders and amplifiers.

2) Calculate the transfer function associated with this system.

3) Calculate its characteristic polynomial.

EXERCISE 3.19.– Jordan normal form

The system described by the state equations:

$$\dot{\mathbf{x}} = \begin{pmatrix} -2 & 1 & 0 & 0 & 0 \\ 0 & -2 & 1 & 0 & 0 \\ 0 & 0 & -2 & 0 & 0 \\ 0 & 0 & 0 & -3 & 1 \\ 0 & 0 & 0 & 0 & -3 \end{pmatrix} \mathbf{x} + \begin{pmatrix} 0 \\ 0 \\ 1 \\ 0 \\ 1 \end{pmatrix} u$$

$$y = \begin{pmatrix} -2 & -1 & 3 & -4 & 7 \end{pmatrix} \mathbf{x} + 2u$$

is in Jordan normal form since its evolution matrix $\mathbf{A}$ is a Jordan matrix. In other words, it is a block diagonal and each block (here there are two) has zeros everywhere, except on the diagonal that contains equal elements and on the opposite diagonal that only contains ones. Moreover, the control matrix only contains ones and zeros that are positioned at the last row of each block.

1) Draw a wiring system for this system using integrators, adders and amplifiers.

2) Calculate its transfer function.

3) Calculate its characteristic polynomial as well as the associated eigenvalues.

3.5. Solutions

Solution to Exercise 3.1 (solution of a continuous-time linear state equation)

Let us take $\mathbf{z}(t) = e^{-\mathbf{A}t}\mathbf{x}(t)$. We have $\mathbf{x}(t) = e^{\mathbf{A}t}\mathbf{z}(t)$ and therefore, by differentiating $\dot{\mathbf{x}}(t) = \mathbf{A}e^{\mathbf{A}t}\mathbf{z}(t) + e^{\mathbf{A}t}\dot{\mathbf{z}}(t)$. The evolution equation $\dot{\mathbf{x}}(t) = \mathbf{A}\mathbf{x}(t) + \mathbf{B}\mathbf{u}(t)$, is transformed into:

$$\mathbf{A}e^{\mathbf{A}t}\mathbf{z}(t) + e^{\mathbf{A}t}\dot{\mathbf{z}}(t) = \mathbf{A}e^{\mathbf{A}t}\mathbf{z}(t) + \mathbf{B}\mathbf{u}(t)$$

and, after simplification:

$$\dot{\mathbf{z}}(t) = e^{-\mathbf{A}t}\mathbf{B}\mathbf{u}(t)$$

After integrating, we obtain:

$$\mathbf{z}(t) = \mathbf{z}(0) + \int_0^t e^{-\mathbf{A}\tau}\mathbf{B}\mathbf{u}(\tau)d\tau$$

Therefore:

$$\begin{aligned}\mathbf{x}(t) &= e^{\mathbf{A}t}\left(\mathbf{z}(0) + \int_0^t e^{-\mathbf{A}\tau}\mathbf{B}\mathbf{u}(\tau)d\tau\right) \\ &= e^{\mathbf{A}t}\mathbf{z}(0) + \int_0^t e^{\mathbf{A}t}e^{-\mathbf{A}\tau}\mathbf{B}\mathbf{u}(\tau)d\tau \\ &= e^{\mathbf{A}t}\mathbf{x}(0) + \int_0^t e^{\mathbf{A}(t-\tau)}\mathbf{B}\mathbf{u}(\tau)d\tau\end{aligned}$$

Solution to Exercise 3.2 (solution of a discrete-time linear state equation)

The proof can easily be obtained by recurrence. First of all, if $k = 0$, the relation is verified. Let us establish that if it is verified for k, then it is also verified for $k + 1$. We have:

$$\begin{aligned}\mathbf{x}(k+1) &= \mathbf{A}\mathbf{x}(k) + \mathbf{B}\mathbf{u}(k) \\ &= \mathbf{A}\left(\mathbf{A}^k\mathbf{x}(0) + \sum_{\ell=0}^{k-1} \mathbf{A}^{k-\ell-1}\mathbf{B}\mathbf{u}(\ell)\right) + \mathbf{B}\mathbf{u}(k) \\ &= \mathbf{A}^{k+1}\mathbf{x}(0) + \sum_{\ell=0}^{k-1} \mathbf{A}^{k-\ell}\mathbf{B}\mathbf{u}(\ell) + \mathbf{B}\mathbf{u}(k) \\ &= \mathbf{A}^{k+1}\mathbf{x}(0) + \sum_{\ell=0}^{k} \mathbf{A}^{k-\ell}\mathbf{B}\mathbf{u}(\ell)\end{aligned}$$

The relation is therefore also verified for $k + 1$.

Solution to Exercise 3.3 (criterion of stability)

1) Let $\mathbf{x}$ be an eigenvector of $\mathbf{A}$ associated with the eigenvalue λ. We will now show that $\mathbf{x}$ is also an eigenvector of $f(\mathbf{A})$ with eigenvalue $f(\lambda)$. First of all, this property is true if $f(\mathbf{A}) = \mathbf{A}^j$. Indeed, since $\mathbf{A}\mathbf{x} = \lambda\mathbf{x}$, we have:

$$\mathbf{A}^j\mathbf{x} = \mathbf{A}^{j-1}.\underbrace{\mathbf{A}\mathbf{x}}_{\lambda\mathbf{x}} = \lambda\mathbf{A}^{j-1}\mathbf{x} = \lambda\mathbf{A}^{j-2}.\underbrace{\mathbf{A}\mathbf{x}}_{\lambda\mathbf{x}} = \lambda^2\mathbf{A}^{j-2}\mathbf{x}$$

$$= \cdots = \lambda^j\mathbf{A}^0\mathbf{x} = \lambda^j\mathbf{x}$$

We, therefore, have the property $f(\mathbf{A}).\mathbf{x} = f(\lambda).\mathbf{x}$ when $f(\mathbf{A}) = \mathbf{A}^j$. Let us assume that this property is true for two polynomials f_1 and f_2 in $\mathbf{A}$, we will now show that it is also true for $f_1 + f_2$ and αf_1. Since it is true for f_1 and f_2, we have:

$$f_1(\mathbf{A}).\mathbf{x} = f_1(\lambda).\mathbf{x}$$

$$f_2(\mathbf{A}).\mathbf{x} = f_2(\lambda).\mathbf{x}$$

and therefore:

$$(f_1(\mathbf{A})+f_2(\mathbf{A})).\mathbf{x} = f_1(\mathbf{A}).\mathbf{x}+f_2(\mathbf{A}).\mathbf{x} = f_1(\lambda).\mathbf{x}+f_2(\lambda).\mathbf{x}$$
$$= (f_1(\lambda)+f_2(\lambda)).\mathbf{x},$$
$$(\alpha.f_1(\mathbf{A})).\mathbf{x} = \alpha.(f_1(\mathbf{A}).\mathbf{x}) = \alpha.f_1(\lambda).\mathbf{x} = (\alpha.f_1(\lambda)).\mathbf{x}$$

By recurrence, we can deduce that the property is true for all functions $f(\mathbf{A})$ that can be generated from the $\mathbf{A}^j$ by compositions of additions and multiplications by a scalar, in other words for the functions f that are polynomials.

2) If spec $(\mathbf{A}) = \{\lambda_1, \ldots, \lambda_n\}$ then:

$$\text{spec}\,(\mathbf{I}+\mathbf{A}) = \{1+\lambda_1, \ldots, 1+\lambda_n\}$$
$$\text{spec}\,(\mathbf{A}^k) = \{\lambda_1^k, \ldots, \lambda_n^k\}$$
$$\text{spec}\,(e^{\mathbf{A}t}) = \{e^{\lambda_1 t}, \ldots, e^{\lambda_n t}\}$$
$$\text{spec}\,(f(\mathbf{A})) = \{f(\lambda_1), \ldots, f(\lambda_n)\} = \{0, \ldots, 0\}$$

Let us note that $f(\mathbf{A})$ is the zero matrix, which corresponds to the Cayley-Hamilton theorem, which states that every square matrix cancels out its characteristic polynomial.

3) Let us take $f(\mathbf{A}) = e^{\mathbf{A}t} = \sum_{j=0}^{\infty} \frac{1}{j!}(\mathbf{A}t)^j$, the theorem of correspondence of the eigenvalues (which we will assume is applicable even for polynomials of infinite degree, i.e. integer series), tells us that the eigenvalues of $e^{\mathbf{A}t}$ are of the form $e^{\lambda_j t}$, where λ_j denotes the j^{ime} eigenvalue of $\mathbf{A}$. However, the

stability of the system is expressed using the condition $\lim_{t\to\infty} e^{\mathbf{A}t} = \mathbf{0}_n$. We have:

$$e^{\mathbf{A}t} \xrightarrow{t\to\infty} \mathbf{0}_n \Rightarrow \forall j \in \{1,\ldots,n\}, e^{\lambda_j t} \xrightarrow{t\to\infty} 0$$
(continuity of the exponential
$$\Leftrightarrow \forall j \in \{1,\ldots,n\}, |e^{(\operatorname{Re}\lambda_j + i\operatorname{Im}\lambda_j)t}| \xrightarrow{t\to\infty} 0$$
and of the eigenvalues)
$$\Leftrightarrow \forall j \in \{1,\ldots,n\}, \underbrace{|e^{(\operatorname{Re}\lambda_j)t}|}_{=e^{(\operatorname{Re}\lambda_j)t}} \cdot \underbrace{|e^{i(\operatorname{Im}\lambda_j)t}|}_{=1} \xrightarrow{t\to\infty} 0$$
$$\Leftrightarrow \forall j \in \{1,\ldots,n\}, \operatorname{Re}(\lambda_j) < 0$$

4) In the discrete-time case, we need to take $f(\mathbf{A}) = \mathbf{A}^k$. A similar reasoning to the continuous-time case gives us:

$$\mathbf{A}^k \xrightarrow{k\to\infty} \mathbf{0}_n \Rightarrow \forall j \in \{1,\ldots,n\}, \lambda_j^k \xrightarrow{k\to\infty} 0 \quad \text{(continuity)}$$
$$\Leftrightarrow \forall j \in \{1,\ldots,n\}, |\underbrace{\left(\rho_j e^{i\theta_j}\right)^k}_{=\rho_j^k e^{ik\theta_j}}| \xrightarrow{k\to\infty} 0 \text{ (polar form)}$$
$$\Leftrightarrow \forall j \in \{1,\ldots,n\}, \rho_j^k \xrightarrow{k\to\infty} 0$$
$$\Leftrightarrow \forall j \in \{1,\ldots,n\}, \rho_j < 1$$

Solution to Exercise 3.4 (Laplace variable)

1) This system can be described by the following equations:

$$\begin{cases} \dot{y}_1 + y_1 = u \\ \dot{y}_2 + 2y_2 = -u \\ y_1 + y_2 = y \end{cases}$$

Thus we need to get rid of the y_1 and y_2. By differentiating the above equations, we obtain:

$$\begin{cases} \ddot{y}_1 + \dot{y}_1 = \dot{u} \\ \ddot{y}_2 + 2\dot{y}_2 = -\dot{u} \\ \dot{y}_1 + \dot{y}_2 = \dot{y} \\ \ddot{y}_1 + \ddot{y}_2 = \ddot{y} \end{cases}$$

We have in total 7 equations with 6 surplus variables: y_1, y_2, $\dot{y}_1, \dot{y}_2, \ddot{y}_1, \ddot{y}_2$. We can eliminate them by a substitution method and thus obtain the differential equation that we were looking for:

$$\ddot{y} + 3\dot{y} + 2y = u$$

2) We have:

$$\begin{cases} \left(\frac{d}{dt} + 1\right)(y_1) = u \\ \left(\frac{d}{dt} + 2\right)(y_2) = -u \\ y_1 + y_2 = y \end{cases}$$

Let us dare the following calculation:

$$\begin{cases} y_1 = \frac{1}{\frac{d}{dt}+1} \cdot u \\ y_2 = -\frac{1}{\frac{d}{dt}+2} \cdot u \end{cases}$$

Whence:

$$y = y_1 + y_2 = \frac{1}{\frac{d}{dt}+1} \cdot u - \frac{1}{\frac{d}{dt}+2} \cdot u$$

$$= \left(\frac{1}{\frac{d}{dt}+1} - \frac{1}{\frac{d}{dt}+2}\right) \cdot u$$

After reducing to the same denominator:

$$y = \left(\frac{\left(\frac{d}{dt}+2\right)-\left(\frac{d}{dt}+1\right)}{\left(\frac{d}{dt}+1\right)\left(\frac{d}{dt}+2\right)} \right) \cdot u = \frac{1}{\frac{d^2}{dt^2}+3\frac{d}{dt}+2} \cdot u$$

Therefore:

$$\left(\frac{d^2}{dt^2} + 3\frac{d}{dt} + 2 \right) \cdot y = u$$

or alternatively:

$$\ddot{y} + 3\dot{y} + 2y = u$$

This reasoning allows us to obtain the expected result using few calculations in an elegant manner. However, this reasoning needs to be placed within a mathematical framework. This is what the Laplace transform brings us. Let us note that the above reasoning could have lead to an incoherence that we can also observe with the (careless) use of the Laplace transform. For instance, the following reasoning is clearly wrong:

$$\dot{y} = \dot{u} \Leftrightarrow \frac{d}{dt}y = \frac{d}{dt}u \Leftrightarrow y = \frac{\frac{d}{dt}}{\frac{d}{dt}}u \Leftrightarrow y = u$$

We clearly have no right to simplify by $\frac{d}{dt}$ since the set of differential operators generated by $\frac{d}{dt}$ is a ring and not a field. Likewise, in the framework of non-linear differential equations (such as $\ddot{y} + y\dot{y} = u$) we quickly reach absurd reasonings.

3) We have:

$$\begin{cases} (s+1) \cdot y_1 = u \\ (s+2) \cdot y_2 = -u \\ y_1 + y_2 = y \end{cases}$$

and, by eliminating y_1 and y_2:

$$y = \frac{1}{s^2 + 3s + 2} \cdot u$$

Solution to Exercise 3.5 (transfer functions of elementary systems)

1) A differentiator is expressed by the differential equation $y = \dot{u}$, i.e. $y(t) = \frac{d}{dt}(u(t))$. Its transfer function is, therefore, $H(s) = s$.

2) An integrator is expressed by the differential equation $\dot{y} = u$, i.e. $\frac{d}{dt}(y(t)) = u(t)$, or alternatively:

$$y(t) = \left(\frac{1}{\frac{d}{dt}}\right) u(t)$$

Its transfer function is, therefore, $H(s) = s^{-1}$.

3) A delay of τ is expressed by the input–output relation $y(t) = u(t-\tau)$. Let us assume that the function $u(t)$ is analytic. An integer series development of $u(t)$ around t_0 gives us:

$$u(t) = \sum_{i=0}^{\infty} \frac{1}{i!} u^{(i)}(t_0) \cdot (t - t_0)^i$$

Let us replace t by $t - \tau$ then t_0 by t, we obtain:

$$u(t - \tau) = \sum_{i=0}^{\infty} \frac{1}{i!} u^{(i)}(t) \cdot (-\tau)^i$$

Thus:

$$y(t) = u(t - \tau) = \sum_{i=0}^{\infty} \frac{1}{i!} u^{(i)}(t) \cdot (-\tau)^i = \left(\sum_{i=0}^{\infty} \frac{1}{i!} \left(-\tau \frac{d}{dt}\right)^i\right)(u(t))$$

$$= e^{-\tau \frac{d}{dt}}(u(t))$$

Thus, the differential operator $e^{-\tau \frac{d}{dt}}$ corresponds to a delay of τ of the signal $u(t)$. The transfer function of the linear system, which generates an output signal identical to the input signal, but delayed by τ, is given by $H(s) = e^{-\tau s}$.

Solution to Exercise 3.6 (transfer function of composite systems)

1) The transfer function of the composite system is $H(s) = H_2(s).H_1(s)$. Indeed, $\hat{y}(s) = H_2(s).H_1(s)\hat{u}(s)$.

2) We have $\hat{y}(s) = \hat{y}_1(s) + \hat{y}_2(s) = H_1(s)\hat{u}(s) + H_2(s)\hat{u}(s) = (H_1(s) + H_2(s))\hat{u}(s)$. Thus, the transfer function of the composite system is $H(s) = H_1(s) + H_2(s)$.

3) Since $\hat{y}(s) = H(s)\hat{e}(s)$ and that $\hat{e}(s) = \hat{u}(s) - \hat{y}(s)$, we have $\hat{y}(s) = H(s)(\hat{u}(s) - \hat{y}(s))$. By isolating $\hat{y}(s)$, we obtain:

$$\hat{y}(s) = \frac{H(s)}{1 + H(s)}\hat{u}(s)$$

The transfer function of the looped system is, therefore, $\frac{H(s)}{1+H(s)}$.

Solution to Exercise 3.7 (transfer matrix)

1) The transfer matrix $\mathbf{G}(s)$ is given by:

$$\mathbf{G}(s) = \mathbf{C}(s\mathbf{I} - \mathbf{A})^{-1}\mathbf{B} + \mathbf{D}$$

$$= \begin{pmatrix} 1 & 2 \\ 1 & 0 \end{pmatrix} \begin{pmatrix} s-1 & -3 \\ -2 & s \end{pmatrix}^{-1} \begin{pmatrix} 1 \\ 1 \end{pmatrix} + \begin{pmatrix} 2 \\ 0 \end{pmatrix}$$

$$= \begin{pmatrix} \frac{2s^2+s-7}{s^2-s-6} \\ \frac{s+3}{s^2-s-6} \end{pmatrix}$$

2) The relation $\hat{\mathbf{y}} = \mathbf{G}(s)\hat{u}$ is written as:

$$(s^2 - s - 6)\hat{\mathbf{y}} = \begin{pmatrix} 2s^2 + s - 7 \\ s + 3 \end{pmatrix} \hat{u}$$

Or alternatively, by substituting s by $\frac{d}{dt}$:

$$\begin{cases} \ddot{y}_1 - \dot{y}_1 - 6y_1 = 2\ddot{u} + \dot{u} - 7u \\ \ddot{y}_2 - \dot{y}_2 - 6y_2 = \dot{u} + 3u \end{cases}$$

Solution to Exercise 3.8 *(matrix block multiplication)*

1) Let us denote by ℓ_M and c_M the number of rows and columns of the matrix M. We have:

$$\ell_{\mathbf{A}_{ik}} = \ell_{\mathbf{C}_{ij}},\ c_{\mathbf{B}_{kj}} = c_{\mathbf{C}_{ij}},\ c_{\mathbf{A}_{ik}} = \ell_{\mathbf{B}_{kj}}$$

i.e.:

$$\ell_{\mathbf{A}_{i1}} = \ell_{\mathbf{A}_{i2}} = \ell_{\mathbf{A}_{i3}} = \ell_{\mathbf{C}_{i1}} = \ell_{\mathbf{C}_{i2}} = \ell_{\mathbf{C}_{i3}},\ i \in \{1,2\}$$
$$c_{\mathbf{B}_{1j}} = c_{\mathbf{B}_{2j}} = c_{\mathbf{B}_{3j}} = c_{\mathbf{C}_{1j}} = c_{\mathbf{C}_{2j}},\ j \in \{1,2,3\}$$
$$c_{\mathbf{A}_{1k}} = c_{\mathbf{A}_{2k}} = \ell_{\mathbf{B}_{k1}} = \ell_{\mathbf{B}_{k2}} = \ell_{\mathbf{B}_{k3}},\ k \in \{1,2,3\}$$

We therefore need $2 + 3 + 3 = 8$ pieces of information on the rows and columns in order to find all the dimensions.

2) A linear resolution of this system of equations gives us the dimensions of Figure 3.10.

Solution to Exercise 3.9 *(change of basis)*

1) By replacing x by Pv, we obtain:

$$\begin{cases} \mathbf{P}\dot{\mathbf{v}} = \mathbf{APv} + \mathbf{Bu} \\ \mathbf{y} = \mathbf{CPv} + \mathbf{Du} \end{cases}$$

i.e.:

$$\begin{cases} \dot{\mathbf{v}} = \mathbf{P}^{-1}\mathbf{APv} + \mathbf{P}^{-1}\mathbf{Bu} \\ \mathbf{y} = \mathbf{CPv} + \mathbf{Du} \end{cases}$$

which is a state representation. Thus, a linear system has as many state representations as there are transfer matrices. Of course, some representations are preferable depending on the desired application.

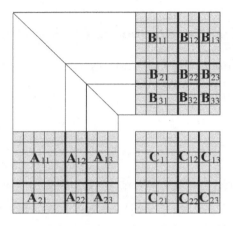

Figure 3.10. *Dimensions of the submatrices involved in the block multiplication*

2) The Jordan normal form decomposition of the evolution matrix is:

$$\underbrace{\begin{pmatrix} 1 & 1 & 1 \\ 2 & 1 & 2 \\ 2 & 1 & 0 \end{pmatrix}^{-1}}_{\mathbf{P}^{-1}} \underbrace{\begin{pmatrix} 4 & -\frac{1}{2} & -\frac{1}{2} \\ 4 & 1 & -1 \\ 4 & -2 & 2 \end{pmatrix}}_{\mathbf{A}} \underbrace{\begin{pmatrix} 1 & 1 & 1 \\ 2 & 1 & 2 \\ 2 & 1 & 0 \end{pmatrix}}_{\mathbf{P}} = \underbrace{\begin{pmatrix} 2 & 1 & 0 \\ 0 & 2 & 0 \\ 0 & 0 & 3 \end{pmatrix}}_{\bar{\mathbf{A}}}$$

Thus, following the basis change relation as seen above, the new state representation is:

$$\begin{cases} \dot{\mathbf{v}} = \begin{pmatrix} 2 & 1 & 0 \\ 0 & 2 & 0 \\ 0 & 0 & 3 \end{pmatrix} \mathbf{v} + \begin{pmatrix} 2 \\ 0 \\ -1 \end{pmatrix} u \\ y = \begin{pmatrix} 6 & 4 & 4 \end{pmatrix} \mathbf{v} \end{cases}$$

3) The transfer function of this system is:

$$\begin{pmatrix} 6 & 4 & 4 \end{pmatrix} \left(\begin{pmatrix} s & 0 & 0 \\ 0 & s & 0 \\ 0 & 0 & s \end{pmatrix} - \begin{pmatrix} 2 & 1 & 0 \\ 0 & 2 & 0 \\ 0 & 0 & 3 \end{pmatrix} \right)^{-1} \begin{pmatrix} 2 \\ 0 \\ -1 \end{pmatrix} = \frac{12}{s-2} - \frac{4}{s-3}$$

Its characteristic polynomial $(s-2)(s-3)$ is of degree 2, whereas the evolution matrix admits $(s-2)^2(s-3)$ as characteristic polynomial. This fact is representative of a pole-zero cancellation in transfer function calculations.

Solution to Exercise 3.10 (change of basis toward a companion matrix)

From the equation $\dot{x} = Ax + bu$ and by taking $v = P^{-1}x$, we obtain $P\dot{v} = APv + bu$. In other words:

$$\dot{v} = P^{-1}APv + P^{-1}bu$$

If we take $\bar{A} = P^{-1}AP$ and $\bar{b} = P^{-1}b$, the system is written as:

$$\dot{v} = \bar{A}v + \bar{b}u$$

Since $P = (b \mid Ab \mid A^2b \mid \ldots \mid A^{n-1}b)$, if e_i denotes the vector that only contains zeros, except a 1 at position i, we have $Pe_i = A^{i-1}b$. Thus:

$$e_i = P^{-1}.A^{i-1}b$$

Therefore, $\bar{b} = P^{-1}b = e_1$. In order to obtain $\bar{A}$, we write:

$$\begin{aligned} \bar{A} &= (\bar{a}_1 \mid \bar{a}_2 \mid \bar{a}_3 \mid \ldots \mid \bar{a}_{n-1} \mid \bar{a}_n) \\ &= P^{-1}AP \\ &= \left(P^{-1}Ab \mid P^{-1}A^2b \mid P^{-1}A^3b \mid \ldots \right. \\ &\quad \left. \mid P^{-1}A^{n-1}b \mid P^{-1}A^nb \right) \\ &= \left(e_2 \mid e_3 \mid e_4 \mid \ldots \mid e_n \mid P^{-1}A^nb \right) \end{aligned}$$

where the $\bar{\mathbf{a}}_i$ denotes the i^{imes} columns of $\bar{\mathbf{A}}$. However, the *Cayley-Hamilton* theorem tells us that every square matrix cancels out its characteristic polynomial:

$$P(s) = s^n + a_{n-1}s^{n-1} + \cdots + a_1 s + a_0$$

Thus $\mathbf{A}^n + a_{n-1}\mathbf{A}^{n-1} + \cdots + a_1\mathbf{A} + a_0\mathbf{I} = \mathbf{0}$, or alternatively:

$$\mathbf{A}^n = -a_{n-1}\mathbf{A}^{n-1} - \cdots - a_1\mathbf{A} - a_0\mathbf{I}$$

By multiplying the left-hand side by $\mathbf{P}^{-1}$ and the right hand side by b, we obtain:

$$\mathbf{P}^{-1}\mathbf{A}^n \mathbf{b} = \mathbf{P}^{-1}\left(-a_{n-1}\mathbf{A}^{n-1}\mathbf{b} + \cdots - a_1\mathbf{A}\mathbf{b} - a_0\mathbf{b}\right)$$

$$= \mathbf{P}^{-1}.\mathbf{P}\begin{pmatrix} -a_0 \\ \vdots \\ -a_{n-1} \end{pmatrix} = \begin{pmatrix} -a_0 \\ \vdots \\ -a_{n-1} \end{pmatrix}$$

In conclusion, after change of basis, we obtain an evolution equation such as:

$$\dot{\mathbf{v}} = \begin{pmatrix} 0 & 0 & 0 & -a_0 \\ 1 & 0 & 0 & -a_1 \\ 0 & \vdots & 0 & \cdots \\ 0 & 0 & 1 & -a_{n-1} \end{pmatrix} \mathbf{v} + \begin{pmatrix} 1 \\ 0 \\ \vdots \\ 0 \end{pmatrix} \mathbf{u}$$

Solution to Exercise 3.11 (pole-zero cancellation)

By the differential method, we have:

$$\dot{y} = \dot{x} + \dot{u} = -x + \dot{u} = -y + u + \dot{u}$$

and therefore $\dot{y} + y = u + \dot{u}$. By Laplace, we have:

$$\begin{cases} s\hat{x} = -\hat{x} \\ \hat{y} = \hat{x} + \hat{u} \end{cases} \Rightarrow \begin{cases} \hat{x} = \frac{0}{s+1} \\ \hat{y} = \hat{x} + \hat{u} \end{cases} \Rightarrow \hat{y} = \frac{0}{s+1} + \hat{u} = \frac{0 + (s+1)u}{s+1}$$

$$\Rightarrow (s+1)\hat{y} = (s+1)\hat{u}$$

and therefore $\dot{y} + y = u + \dot{u}$. Let us note that we have forbidden ourselves from using pole-zero cancellations. Strictly speaking, we should have taken into account the initial conditions, which are not done in practice by control engineers when calculating transfer functions. Often, we are allowed to perform pole-zero cancellations when these are stable. Thus, we could write:

$$\frac{(s+a)(s+3)}{(s+a)(s+2)} = \frac{s+3}{s+2}$$

if $a > 0$. In our example $\dot{y} + y = u + \dot{u}$, if we are allowed to use this pole-zero cancellation, we obtain:

$$\hat{y} = \frac{s+1}{s+1}\hat{u} = \hat{u}$$

We have simplified the pole -1 (in the denominator) with the zero -1 (in the numerator). Mathematically, this is not correct, but in practice, it is. Indeed, we have the following solution for the differential equation:

$$y(t) = \alpha e^{-t} + u(t)$$

in other words, once the steady state has been reached, we will have $y(t) = u(t)$. In the case of an unstable zero, this is no longer the case given the fact that the dependence on the initial conditions remains.

Solution to Exercise 3.12 (state equations of a wiring system)

1) We have:

$$\begin{cases} \dot{x}_1 = -x_2 + u - 3x_1 + y \\ \dot{x}_2 = x_1 + x_2 \\ y = 2x_1 + x_2 \end{cases} \Leftrightarrow \begin{cases} \dot{x}_1 = u - x_1 \\ \dot{x}_2 = x_1 + x_2 \\ y = 2x_1 + x_2 \end{cases}$$

Thus:
$$\begin{cases} \dot{\mathbf{x}} = \begin{pmatrix} -1 & 0 \\ 1 & 1 \end{pmatrix} \mathbf{x} + \begin{pmatrix} 1 \\ 0 \end{pmatrix} u \\ y = \begin{pmatrix} 2 & 1 \end{pmatrix} \mathbf{x} \end{cases}$$

2) We have $P(s) = (s+1)(s-1)$. The system is unstable, since the pole $+1$ has positive real part.

3) In the Laplace domain, the state equations are written as:
$$\begin{cases} s\hat{x}_1 = \hat{u} - \hat{x}_1 \\ s\hat{x}_2 = \hat{x}_1 + \hat{x}_2 \\ \hat{y} = 2\hat{x}_1 + \hat{x}_2 \end{cases}$$

And thus:
$$\begin{cases} \hat{x}_1 = \frac{\hat{u}}{s+1} \\ \hat{x}_2 = \frac{\hat{x}_1}{s-1} = \frac{1}{s-1} \cdot \frac{\hat{u}}{s+1} \\ \hat{y} = 2\frac{\hat{u}}{s+1} + \frac{1}{s-1} \cdot \frac{\hat{u}}{s+1} = \frac{2s-1}{(s+1)(s-1)} \hat{u} \end{cases}$$

The transfer function is therefore:
$$H(s) = \frac{2s-1}{(s+1)(s-1)} = \frac{2s-1}{s^2 - 1}$$

Solution to Exercise 3.13 (combination of systems)

1) The state equations of the system S_a are:
$$S_a : \begin{cases} \dot{\mathbf{x}} = \begin{pmatrix} 0 & 0 \\ 1 & 2 \end{pmatrix} \mathbf{x} + \begin{pmatrix} 1 \\ 0 \end{pmatrix} u \\ y = \begin{pmatrix} 0 & 1 \end{pmatrix} \mathbf{x} \end{cases}$$

diagram, we can directly write the state equations of this system:

$$\begin{cases} \begin{pmatrix} \dot{x}_1 \\ \dot{x}_2 \\ \dot{x}_3 \end{pmatrix} = \begin{pmatrix} 0 & 1 & 0 \\ 0 & 0 & 1 \\ -a_0 & -a_1 & -a_2 \end{pmatrix} \begin{pmatrix} x_1 \\ x_2 \\ x_3 \end{pmatrix} + \begin{pmatrix} 0 \\ 0 \\ 1 \end{pmatrix} u \\ y = \begin{pmatrix} b_0 & b_1 & b_2 \end{pmatrix} \begin{pmatrix} x_1 \\ x_2 \\ x_3 \end{pmatrix} \end{cases}$$

This reasoning can be applied in order to find the state representation of any monovariate linear system of any order n. This particular form for the state representation, which involves the coefficients of the transfer function in the matrices is called *canonical form of control*. Thus, generally, in order to obtain the canonical form of control equivalent to a given monovariate linear system, we have to calculate its transfer function in its developed form. Then, we can immediately write its canonical form of control.

Solution to Exercise 3.17 (canonical observation form)

1) Let us isolate $\dddot{y}$ (which corresponds to the term differentiated the most amount of times). We obtain:

$$\dddot{y} = b_2 \ddot{u} - a_2 \ddot{y} + b_1 \dot{u} - a_1 \dot{y} + b_0 u - a_0 y$$

By integrating, we can remove the differentials:

$$y = \int \int \int (b_2 \ddot{u} - a_2 \ddot{y} + b_1 \dot{u} - a_1 \dot{y} + b_0 u - a_0 y)$$

2) We have:

$$y = \int \{ \, b_2 u - a_2 y + \underbrace{\int [\, b_1 u - a_1 y + \underbrace{\int (b_0 u - a_0 y)}_{\dot{x}_1} \,]}_{\dot{x}_2} \, \}$$

$$\underbrace{\phantom{y = \int \{ \, b_2 u - a_2 y + \int [\, b_1 u - a_1 y + \int (b_0 u - a_0 y) \,] \, \}}}_{\dot{x}_3}$$

By defining x_1, x_2, x_3 as the value of each of these three integrals, i.e.:

$$\begin{cases} x_1 = \int (b_0 u - a_0 y) \\ x_2 = \int (b_1 u - a_1 y + x_1) \\ x_3 = \int (b_2 u - a_2 y + x_2) \end{cases}$$

from this we deduce the wiring of Figure 3.13. Let us note that this diagram is strangely reminiscent of the one obtained in Exercise 3.16, supposed to represent the same system. We move from one to the other by changing the direction of the arrows, replacing the adders with soldered joints and the joints with adders.

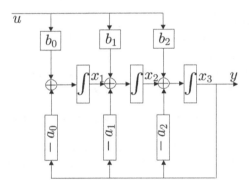

Figure 3.13. *Canonical observation form for a system of order 3*

3) From this diagram, we can directly deduce the state equations of the system:

$$\begin{cases} \begin{pmatrix} \dot{x}_1 \\ \dot{x}_2 \\ \dot{x}_3 \end{pmatrix} = \begin{pmatrix} 0 & 0 & -a_0 \\ 1 & 0 & -a_1 \\ 0 & 1 & -a_2 \end{pmatrix} \begin{pmatrix} x_1 \\ x_2 \\ x_3 \end{pmatrix} + \begin{pmatrix} b_0 \\ b_1 \\ b_2 \end{pmatrix} u \\ y = \begin{pmatrix} 0 & 0 & 1 \end{pmatrix} \begin{pmatrix} x_1 \\ x_2 \\ x_3 \end{pmatrix} \end{cases}$$

This particular form for the state representation is called *canonical observation form*.

4) Let us note that the transformation $\mathbf{A} \to \mathbf{A}^T$, $\mathbf{B} \to \mathbf{C}^T, \mathbf{C} \to \mathbf{B}^T$, gives us the canonical form of control (refer to Exercise 3.16). Let us now try to explain why this transposition, which makes us move from the canonical form of control to the canonical observation form and vice-versa, does not change the input–output behavior of the system. For this, let us consider the system Σ whose state matrices are $(\mathbf{A}, \mathbf{B}, \mathbf{C}, \mathbf{D})$ and the system Σ' whose state matrices are $(\mathbf{A}' = \mathbf{A}^T, \mathbf{B}' = \mathbf{C}^T, \mathbf{C}' = \mathbf{B}^T, \mathbf{D}' = \mathbf{D}^T)$. The transfer matrix associated with Σ is:

$$\mathbf{G}(s) = \mathbf{C}(s\mathbf{I} - \mathbf{A})^{-1}\mathbf{B} + \mathbf{D}$$

The transfer matrix associated with Σ' is:

$$\begin{aligned}\mathbf{G}'(s) &= \mathbf{C}'(s\mathbf{I} - \mathbf{A}')^{-1}\mathbf{B}' + \mathbf{D}' = \mathbf{B}^T\left(s\mathbf{I} - \mathbf{A}^T\right)^{-1}\mathbf{C}^T + \mathbf{D}^T \\ &= \mathbf{B}^T\left((s\mathbf{I} - \mathbf{A})^{-1}\right)^T \mathbf{C}^T + \mathbf{D}^T = \left(\mathbf{C}(s\mathbf{I} - \mathbf{A})^{-1}\mathbf{B}\right)^T + \mathbf{D}^T \\ &= \left(\mathbf{C}(s\mathbf{I} - \mathbf{A})^{-1}\mathbf{B} + \mathbf{D}\right)^T = \mathbf{G}^T(s)\end{aligned}$$

However, here $\mathbf{G}(s)$ is a scalar and therefore $\mathbf{G}(s) = \mathbf{G}^T(s)$. Thus, the transformation $\mathbf{A} \to \mathbf{A}^T, \mathbf{B} \to \mathbf{C}^T, \mathbf{C} \to \mathbf{B}^T, \mathbf{D} \to \mathbf{D}^T$ does not change the transfer function of the system if it has a single input and a single output.

Solution to Exercise 3.18 (modal form)

1) The wiring of the system is given in Figure 3.14.

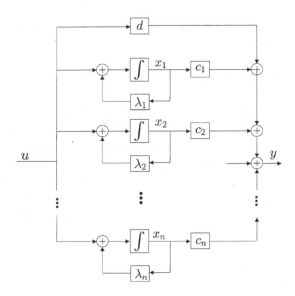

Figure 3.14. *Linear system in modal form*

2) The transfer function of the system is given by:

$$G(s) = \mathbf{C}(s\mathbf{I} - \mathbf{A})^{-1}\mathbf{B} + d$$

$$= \begin{pmatrix} c_1 & \cdots & c_n \end{pmatrix} \begin{pmatrix} s - \lambda_1 & 0 & \cdots \\ & & \\ 0 & \ddots & 0 \\ & & \\ \cdots & 0 & s - \lambda_n \end{pmatrix}^{-1} \begin{pmatrix} 1 \\ 1 \\ \cdots \\ 1 \end{pmatrix} + d$$

$$= \begin{pmatrix} c_1 & \cdots & c_n \end{pmatrix} \begin{pmatrix} \frac{1}{s-\lambda_1} & 0 & \cdots & 0 \\ 0 & \frac{1}{s-\lambda_2} & 0 & \cdots \\ \cdots & \cdots & \ddots & \cdots \\ 0 & \cdots & 0 & \frac{1}{s-\lambda_n} \end{pmatrix} \begin{pmatrix} 1 \\ 1 \\ \cdots \\ 1 \end{pmatrix} + d$$

$$= \frac{c_1}{s - \lambda_1} + \frac{c_2}{s - \lambda_2} + \cdots + \frac{c_n}{s - \lambda_n} + d$$

3) Its characteristic polynomial is given by $\det(s\mathbf{I} - \mathbf{A})$, i.e.:

$$\det \begin{pmatrix} s - \lambda_1 & 0 & \cdots & 0 \\ 0 & s - \lambda_2 & 0 & \cdots \\ \cdots & \cdots & \cdots & \cdots \\ 0 & \cdots & 0 & s - \lambda_n \end{pmatrix} = (s - \lambda_1)(s - \lambda_2)\ldots(s - \lambda_n)$$

Its roots are the λ_i that are also the eigenvalues of $\mathbf{A}$.

Solution to Exercise 3.19 (Jordan normal form)

1) A wiring for this system is given in Figure 3.15.

2) Its transfer function is given by:

$$G(s) = 2 + \frac{3}{s+2} - \frac{1}{(s+2)^2} - \frac{2}{(s+2)^3} + \frac{7}{s+3} - \frac{4}{(s+3)^2}$$

3) Its characteristic polynomial is:

$$P(s) = (s+2)^3(s+3)^2$$

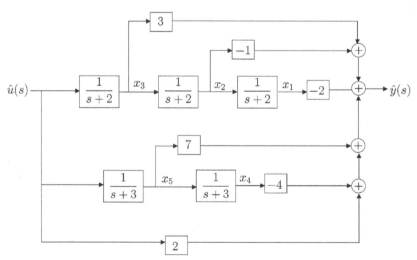

Figure 3.15. *System in its Jordan normal form*

4

Linear Control

In this chapter, we will study the design of controllers for systems given by linear state equations of the form:

$$\begin{cases} \dot{x} = Ax + Bu \\ y = Cx \end{cases}$$

We will show in the following chapter that around very particular points of the state space, called *operating points*, many nonlinear systems genuinely behave like linear systems. The techniques developed in this chapter will then be used for the control of nonlinear systems. Let us denote by m, n, p the respective dimensions of the vectors u, x and y. Recall that **A** is called *evolution matrix*, **B** is the *control matrix* and **C** is the *observation matrix*. We have assumed here, in the interest of simplification, that the direct matrix D involved in the observation equation was nil. In the case where such a direct matrix exists, we can remove it with a simple loop as shown in Exercise 4.1.

After having defined the fundamental concepts of controllability and observability, we will propose two approaches for the design of controllers. First of all, we will assume that the state x is accessible on demand. Even though this hypothesis is generally not verified, it will allow us to establish the principles of the *pole placement method*. In the second phase, we will no longer assume that the state is accessible. We will then have to develop *state estimators* capable of approximating the state vector in order to be able to employ the tools developed in the first phase. The readers should consult Kailath [KAI 80] in order to have a wider view of the methods used for the control of linear systems. A complete and pedagogic course accompanied by numerous exercises can be found in the books of Rivoire and Ferrier [RIV 89].

4.1. Controllability and observability

There are multiple equivalent definitions for the controllability and observability of linear systems. A simple definition is the following.

DEFINITION.– The linear system:

$$\begin{cases} \dot{\mathbf{x}} = \mathbf{A}\mathbf{x} + \mathbf{B}\mathbf{u} \\ \mathbf{y} = \mathbf{C}\mathbf{x} \end{cases}$$

is said to be *controllable* if, for every pair of state vectors $(\mathbf{x}_0, \mathbf{x}_1)$, we can find a time t_1 and a control $\mathbf{u}(t)$, $t \in [0, t_1]$, such that the system, initialized in $\mathbf{x}_0$, reaches the state $\mathbf{x}_1$, at time t_1. It is *observable* is the knowledge of $\mathbf{y}(t)$ and of $\mathbf{u}(t)$ for all $t \in \mathbb{R}$ allows us to determine, in a unique manner, the state $\mathbf{x}(t)$.

CRITERION OF CONTROLLABILITY.– The system is controllable if and only if:

$$\text{rank}\underbrace{\left(\mathbf{B} \mid \mathbf{AB} \mid \mathbf{A}^2\mathbf{B} \mid \ldots \mid \mathbf{A}^{n-1}\mathbf{B}\right)}_{\Gamma_{\text{con}}} = n$$

where n is the dimension of $\mathbf{x}$. In other words, the matrix Γ_{con}, called the *controllability matrix*, obtained by juxtaposing the

n matrices $B, AB, \ldots, A^{n-1}B$ next to one another, has to be of full rank in order for the system to be controllable. This criterion is proved in Exercise 4.5.

CRITERION OF OBSERVABILITY.– The linear system is *observable* if:

$$\mathrm{rank}\underbrace{\begin{pmatrix} C \\ CA \\ \vdots \\ CA^{n-1} \end{pmatrix}}_{\Gamma_{\mathrm{obs}}} = n$$

in other words the matrix, called the *observability matrix* Γ_{obs}, obtained by placing the n matrices $C, CA, \ldots, CA^{n-1}$ below one another, is of full rank. The proof of this criterion is discussed in Exercise 4.6.

The above definitions as well as the criteria are also valid for discrete-time linear systems (see Exercise 4.4).

4.2. State feedback control

Let us consider the system which is assumed to be controllable $\dot{x} = Ax + Bu$, and we are looking to find a controller for this system of the form $u = w - Kx$, where w is the new input. This leads to the assumption that x is accessible on demand, which is normally not the case. We will see further on how to get rid of this awkward hypothesis. The state equations of the looped system are written as:

$$\dot{x} = Ax + B(w - Kx) = (A - BK)x + Bw$$

It is reasonable to choose the control matrix K so as to impose the poles of the looped system. This problem is equivalent to imposing the characteristic polynomial of the system. Let $P_{\text{con}}(s)$ be the desired polynomial, that we will of course assume to be of degree n. We need to solve the polynomial equation:

$$\det(s\mathbf{I} - \mathbf{A} + \mathbf{BK}) = P_{\text{con}}(s)$$

referred to as *pole placement*. This equation can be translated into n scalar equations. Let us indeed recall that two monic polynomials of degree n $s^n + a_{n-1}s + \cdots + a_0$ and $s^n + b_{n-1}s + \cdots + b_0$ are equal if and only if their coefficients are all equal, i.e. if $a_{n-1} = b_{n-1}, \ldots, a_0 = b_0$. Our system of n equations has $m.n$ unknowns which are the coefficients k_{ij}, $i \in \{1, \ldots, m\}$, $j \in \{1, \ldots, n\}$. In fact, a single solution matrix $\mathbf{K}$ is sufficient. We may, therefore, fix $(m-1)$ elements of $\mathbf{K}$ so that we are left with only n unknowns. However, the obtained system is not always linear. The `ppol` instruction in SCILAB or `place` in MATLAB allows us to solve the pole placement equation.

4.3. Output feedback control

In this section, we are once again looking to stabilize the system:

$$\begin{cases} \dot{\mathbf{x}} = \mathbf{Ax} + \mathbf{Bu} \\ \mathbf{y} = \mathbf{Cx} \end{cases}$$

but this time, the state x of the system is is no longer assumed to be accessible on demand. The controller, which will be used, is represented in Figure 4.1.

Only the setpoint w and the output of the system y can be used by the controller. The unknowns of the controller are the matrices $\mathbf{K}$, $\mathbf{L}$ and $\mathbf{H}$. Let us attempt to explain the structure of this controller. First of all, in order to estimate the state x necessary for computing the control u, we integrate a *simulator* of our system into the controller. The simulator is a copy of the system and its state vector is denoted by $\hat{x}$. The error ε_y between the output of the simulator $\hat{y}$ and the output of the system y allows us to correct, by using a correction matrix L, the evolution of the estimated state $\hat{x}$. The corrected simulator is referred to as the *Luenberger observer*. We can then apply a state feedback technique that will be carried out using the matrix K. For the calculation of K, we may use the pole placement method described earlier, which consists of solving $\det(s\mathbf{I} - \mathbf{A} + \mathbf{BK}) = P_{\text{con}}(s)$, where $P_{\text{con}}(s)$ is the characteristic polynomial of degree n chosen for the control dynamics. For the calculation of the correction matrix L, we will try to guarantee a convergence of the error $\hat{x} - x$ to 0. The role of the matrix H (square matrix placed right after the setpoint vector w), called a *precompensator*, is to match the components of the setpoint w with certain state variables chosen beforehand. We will also discuss how to choose this matrix H in order to be able to perform this association.

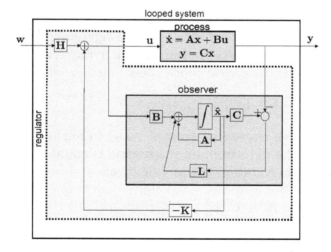

Figure 4.1. *Principle of an output feedback controller*

In order to calculate L, let us extract, from the controlled system of the figure, the subsystem with input u formed from the system to be controlled and its observer (see Figure 4.2).

The state equations that describe the system are:

$$\begin{cases} \dot{x} = Ax + Bu \\ \frac{d}{dt}\hat{x} = A\hat{x} + Bu - L(C\hat{x} - Cx) \end{cases}$$

where the state vector is $(x, \hat{x})$. Let us create, only by thought, the quantity $\varepsilon_x = \hat{x} - x$, as represented in Figure 4.2. By subtracting the two evolution equations, we obtain:

$$\frac{d}{dt}(\hat{x} - x) = A\hat{x} + Bu - L(C\hat{x} - Cx) - Ax - Bu$$
$$= A(\hat{x} - x) - LC(\hat{x} - x)$$

Thus, ε_x obeys the differential equation:

$$\dot{\varepsilon}_x = (A - LC)\varepsilon_x$$

in which the control u is not involved. The estimation error ε_x on the state tends toward zero if all the eigenvalues of $A - LC$ have negative real parts. To impose the dynamics of the error (in other words, its convergence speed) means solving:

$$\det(sI - A + LC) = P_{\text{obs}}(s)$$

where $P_{\text{obs}}(s)$ is chosen as desired, so as to have the required poles. Since the determinant of a matrix is equal to that of its transpose, this equation is equivalent to:

$$\det\left(sI - A^T + C^T L^T\right) = P_{\text{obs}}(s)$$

We obtain a pole placement-type equation.

PRECOMPENSATOR.– The precompensator H allows us to associate the setpoints (components of w) with certain values of the state variables that we are looking at. The choice of these state variables is done by using a setpoint matrix E, as illustrated in Exercise 4.11.

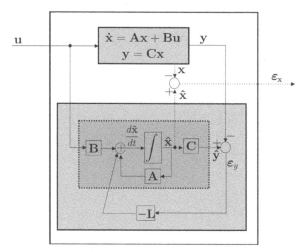

Figure 4.2. *Diagram of the output error between the observer and system*

4.4. Summary

The algorithm of the table below summarizes the method that allows us to calculate an output feedback controller with precompensator.

Algorithm REGULKLH(in: $\mathbf{A}, \mathbf{B}, \mathbf{C}, \mathbf{E}, \mathbf{p}_{con}, \mathbf{p}_{obs}$; out: $\mathcal{R}$)
1 $\mathbf{K} := \text{PLACE}(\mathbf{A}, \mathbf{B}, \mathbf{p}_{con})$;
2 $\mathbf{L} := \text{PLACE}\left(\mathbf{A}^T, \mathbf{C}^T, \mathbf{p}_{obs}\right)^T$;
3 $\mathbf{H} := -\left(\mathbf{E}\left(\mathbf{A} - \mathbf{B}\mathbf{K}\right)^{-1}\mathbf{B}\right)^{-1}$;
4 $\mathcal{R} := \begin{cases} \frac{d}{dt}\hat{\mathbf{x}} = (\mathbf{A} - \mathbf{B}\mathbf{K} - \mathbf{L}\mathbf{C})\,\hat{\mathbf{x}} + \begin{pmatrix} \mathbf{B}\mathbf{H} & \mathbf{L} \end{pmatrix} \begin{pmatrix} \mathbf{w} \\ \mathbf{y} \end{pmatrix} \\ \mathbf{u} = -\mathbf{K}\,\hat{\mathbf{x}} + \begin{pmatrix} \mathbf{H} & 0 \end{pmatrix} \begin{pmatrix} \mathbf{w} \\ \mathbf{y} \end{pmatrix} \end{cases}$

The associated MATLAB function (see the function RegulKLH.m) is given below.

```
function [Ar,Br,Cr,Dr]=RegulKLH(A,B,C,E,pcom,pobs);
    K=place(A,B,pcom)
    L=place(A',C',pobs)'
    H=-inv(E*inv(A-B*K)*B)
    Ar=A-B*K-L*C
    Br=[B*H L]
    Cr=-K
    Dr=[H,0*B'*C']
end;
```

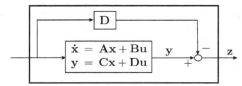

Figure 4.3. *Loop allowing the removal of the direct matrix* **D**

4.5. Exercises

EXERCISE 4.1.– Removing the direct matrix

Let us consider the system:

$$\begin{cases} \dot{x} = Ax + Bu \\ y = Cx + Du \end{cases}$$

and create the new output $z = y - Du$, as represented in Figure 4.3. What do the state equations of the system become?

EXERCISE 4.2.– Non-observable states, non-controllable states

Let us consider the system:

$$\begin{cases} \dot{\mathbf{x}}(t) = \begin{pmatrix} -1 & 1 & 0 & 0 \\ 0 & 1 & 0 & 0 \\ 1 & 1 & 1 & 1 \\ 0 & 1 & 0 & 1 \end{pmatrix} \mathbf{x}(t) + \begin{pmatrix} 1 \\ 0 \\ 1 \\ 0 \end{pmatrix} u(t) \\ \mathbf{y}(t) = \begin{pmatrix} 1 & 1 & 0 & 0 \end{pmatrix} \mathbf{x}(t) + 1 \, u(t) \end{cases}$$

1) Calculate its transfer function.

2) Is this system stable?

3) Draw the associated wiring system and deduce from this the non-observable and non-controllable states.

4) The poles of the system (the eigenvalues of the evolution matrix) are composed of transmission poles (poles of the transfer function) and hidden modes. Give the hidden modes.

EXERCISE 4.3.– Using the controllability and observability criteria

Let us consider the system:

$$\begin{cases} \dot{\mathbf{x}} = \begin{pmatrix} 1 & 1 & 0 \\ 0 & 1 & 0 \\ 0 & 0 & 1 \end{pmatrix} \mathbf{x} + \begin{pmatrix} 0 & 0 \\ 1 & 0 \\ 1 & a \end{pmatrix} \mathbf{u} \\ \mathbf{y} = \begin{pmatrix} 1 & 1 & b \\ 0 & 1 & 0 \end{pmatrix} \mathbf{x} \end{cases}$$

1) For which values of a is the system controllable?

2) For which values of b is the system observable?

EXERCISE 4.4.– **Proof of the controllability criterion in discrete time**

Prove the criterion of controllability in discrete time. This theorem states that the system $\mathbf{x}(k+1) = \mathbf{A}\mathbf{x}(k) + \mathbf{B}\mathbf{u}(k)$ is controllable if and only if:

$$\text{rank}\underbrace{\left(\mathbf{B} \mid \mathbf{AB} \mid \mathbf{A}^2\mathbf{B} \mid \ldots \mid \mathbf{A}^{n-1}\mathbf{B}\right)}_{\Gamma_{\text{con}}} = n$$

where n is the dimension of $\mathbf{x}$.

1) Show that:

$$\mathbf{x}(n) = \mathbf{A}^n\mathbf{x}(0) + \left(\mathbf{B} \mid \mathbf{AB} \mid \ldots \mid \mathbf{A}^{n-1}\mathbf{B}\right)\begin{pmatrix}\mathbf{u}(n-1) \\ \vdots \\ \mathbf{u}(1) \\ \mathbf{u}(0)\end{pmatrix}$$

2) Show that if the matrix $\left(\mathbf{B} \mid \mathbf{AB} \mid \ldots \mid \mathbf{A}^{n-1}\mathbf{B}\right)$ is of full rank then, for every initial vector $\mathbf{x}_0$, for every target vector $\mathbf{x}_n$, we can find a control $\mathbf{u}(0), \mathbf{u}(1), \ldots, \mathbf{u}(n-1)$ such that the system initialized in $\mathbf{x}_0$ reaches $\mathbf{x}_n$, at time $k = n$.

EXERCISE 4.5.– **Proof of the controllability criterion in continuous time**

The objective of this exercise is to prove the criterion of controllability in continuous time. This criterion states that the system $\dot{\mathbf{x}} = \mathbf{A}\mathbf{x} + \mathbf{B}\mathbf{u}$ is controllable if and only if:

$$\text{rank}\underbrace{\left(\mathbf{B} \mid \mathbf{AB} \mid \mathbf{A}^2\mathbf{B} \mid \ldots \mid \mathbf{A}^{n-1}\mathbf{B}\right)}_{\Gamma_{\text{con}}} = n$$

where n is the dimension of $\mathbf{x}$.

1) We assume that:

$$\text{rank}\underbrace{\left(\mathbf{B} \mid \mathbf{AB} \mid \mathbf{A}^2\mathbf{B} \mid \ldots \mid \mathbf{A}^{n-1}\mathbf{B}\right)}_{\Gamma_{\text{con}}} < n$$

where $n = \dim \mathbf{x}$ and Γ_{con} is the controllability matrix. Let $\mathbf{z}$ be a non-zero vector such that $\mathbf{z}^T.\Gamma_{\text{con}} = \mathbf{0}$. By using the solution of the state equation:

$$\mathbf{x}(t) = e^{\mathbf{A}t}\mathbf{x}(0) + \int_0^t e^{\mathbf{A}(t-\tau)}\mathbf{B}\mathbf{u}(\tau)d\tau$$

show that the control $\mathbf{u}$ cannot influence the value $\mathbf{z}^T\mathbf{x}$. Deduce from this that the system is not controllable.

2) Show that if $\text{rank}(\Gamma_{\text{con}}) = n$, for every pair $(\mathbf{x}(0), \mathbf{x}(t_1))$, $t_1 > 0$, there is a polynomial control $\mathbf{u}(t), t \in [0, t_1]$ which leads the system from state $\mathbf{x}(0)$ to state $\mathbf{x}(t_1)$. Here, we will limit ourselves to $t_1 = 1$, knowing that the main principle of the proof remains valid for any $t_1 > 0$.

EXERCISE 4.6.– Proof of the observability criterion

Consider the continuous-time linear system:

$$\dot{\mathbf{x}} = \mathbf{A}\mathbf{x} + \mathbf{B}\mathbf{u}$$

$$\mathbf{y} = \mathbf{C}\mathbf{x}$$

1) Show that:

$$\begin{pmatrix} \mathbf{y} \\ \dot{\mathbf{y}} \\ \ddot{\mathbf{y}} \\ \vdots \\ \mathbf{y}^{(n-1)} \end{pmatrix} = \begin{pmatrix} \mathbf{C} \\ \mathbf{CA} \\ \mathbf{CA}^2 \\ \vdots \\ \mathbf{CA}^{n-1} \end{pmatrix} \mathbf{x} + \begin{pmatrix} 0 & 0 & 0 & 0 \\ \mathbf{CB} & 0 & 0 & 0 \\ \mathbf{CAB} & \mathbf{CB} & 0 & \\ \vdots & & \ddots & \ddots \\ \mathbf{CA}^{n-2}\mathbf{B} & \cdots & \mathbf{CAB} & \mathbf{CB} \end{pmatrix} \begin{pmatrix} \mathbf{u} \\ \dot{\mathbf{u}} \\ \ddot{\mathbf{u}} \\ \vdots \\ \mathbf{u}^{(n-2)} \end{pmatrix}$$

2) Deduce from this that if the observability matrix:

$$\Gamma_{obs} = \begin{pmatrix} C \\ CA \\ \vdots \\ CA^{n-1} \end{pmatrix}$$

is of full rank, then we can express the state x as a linear function of the quantities $u, y, \dot{u}, \dot{y}, \ldots, u^{(n-2)}, y^{(n-2)}, y^{(n-1)}$.

EXERCISE 4.7.– Kalman decomposition

A linear system can always be decomposed, after a suitable change of basis, into four subsystems S_1, S_2, S_3, S_4 where S_1 is controllable and observable, S_2 is non-controllable and observable, S_3 is controllable and non-observable and S_4 is neither controllable nor observable. The dependencies between the subsystems can be summarized in Figure 4.4.

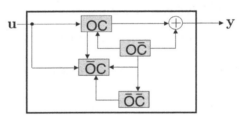

Figure 4.4. *Principle of the Kalman decomposition ; C: controllable, O: observable, $\overline{C}$: non-controllable, $\overline{O}$: non-observable*

Let us note that, in the figure, there is no path (respecting the direction of the arrows) leading from the input u to a non-controllable system. Similarly, there is no path leading from a non-observable system to y. We consider the system described by the state equation:

$$\begin{cases} \dot{\mathbf{x}}(t) = \begin{pmatrix} \mathbf{A}_{11} & \mathbf{A}_{12} & 0 & 0 \\ 0 & \mathbf{A}_{22} & 0 & 0 \\ \mathbf{A}_{31} & \mathbf{A}_{32} & \mathbf{A}_{33} & \mathbf{A}_{34} \\ 0 & \mathbf{A}_{42} & 0 & \mathbf{A}_{44} \end{pmatrix} \mathbf{x}(t) + \begin{pmatrix} \mathbf{B}_1 \\ 0 \\ \mathbf{B}_3 \\ 0 \end{pmatrix} \mathbf{u}(t) \\ \mathbf{y}(t) = \begin{pmatrix} \mathbf{C}_1 & \mathbf{C}_2 & 0 & 0 \end{pmatrix} \mathbf{x}(t) + (\mathbf{D}) \ \mathbf{u}(t) \end{cases}$$

Draw a wiring diagram of the system. Show a decomposition in subsystems $\mathcal{S}_i$ corresponding to the Kalman decomposition.

EXERCISE 4.8.– Resolution of the pole placement equation

We will illustrate here the resolution of the pole placement equation when the system only has a single input. We consider the system:

$$\dot{\mathbf{x}} = \begin{pmatrix} 1 & 2 \\ 3 & 4 \end{pmatrix} \mathbf{x} + \begin{pmatrix} 1 \\ 2 \end{pmatrix} u$$

that we are looking to stabilize by state feedback of the form: $u = w - \mathbf{Kx}$, with: $\mathbf{K} = (k_1 \ k_2)$. Calculate $\mathbf{K}$ so that this characteristic polynomial $P_{\text{con}}(s)$ of the closed-loop system has the roots -1 and -1.

EXERCISE 4.9.– Output feedback of a scalar system

Let us consider the following state equation:

$$\begin{cases} \dot{x} = 3x + 2u \\ y = 4x \end{cases}$$

1) Propose an output feedback controller that puts all the poles in -1 and such that the setpoint variable corresponding to x (in other words if we fix the setpoint at $\bar{w}$, we want the state x to converge toward $\bar{w}$).

2) Give the state equations of the looped system. What are the poles of the looped system?

EXERCISE 4.10.– Separation principle

Let us consider a system looped by a pole placement method, as represented in Figure 4.5.

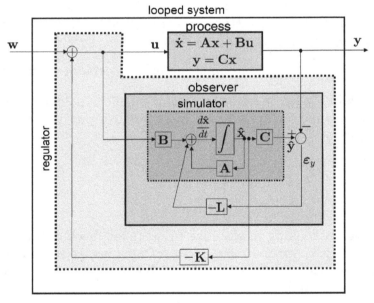

Figure 4.5. *Output feedback controller*

1) Give the state equations of the controller.

2) Find the state equations of the looped system. We will take as state vector the vector $(\mathbf{x}^T \, \hat{\mathbf{x}}^T)^T$.

3) Let us take $\varepsilon_x = \hat{\mathbf{x}} - \mathbf{x}$ and take as new state vector $(\mathbf{x}^T \, \varepsilon_x^T)^T$. Give the associated state representation. Show that ε_x is not controllable.

4) Show that the characteristic polynomial of the looped system is given by:

$$P(s) = \det(s\mathbf{I} - \mathbf{A} + \mathbf{BK}) \cdot \det(s\mathbf{I} - \mathbf{A} + \mathbf{LC})$$

What can be deduced from this about the relationship between the poles of this system and the ones we have placed?

EXERCISE 4.11.– Choosing the precompensator

Let us consider once again the looped linear system of Figure 4.5. We will assume that this system is stable. The setpoint $\mathbf{w}(t)$ is a constant $\bar{\mathbf{w}}$.

1) Toward what values $\bar{\mathbf{x}}$ and $\bar{\varepsilon}_x$ do $\mathbf{x}$ and ε_x converge once the point of equilibrium has been reached?

2) We call *setpoint variables*, $\mathbf{x}_c$ a set of m state variables where $m = \dim(\mathbf{w}) = \dim(\mathbf{u}))$ for which we would like that, for constant $\mathbf{w} = \bar{\mathbf{w}}$, $\mathbf{x}_c$ converges toward $\bar{\mathbf{x}}_c = \bar{\mathbf{w}}$. Let us assume that these variables may be obtained by a linear combination of the components of $\mathbf{x}$ through a relation of type:

$$\mathbf{x}_c = \mathbf{E}\mathbf{x}$$

where $\mathbf{E}$ is an $m \times n$, matrix, referred to as *setpoint matrix*. Let us assume that the state is $\mathbf{x} = (x_1, x_2, x_3)^T$ and that we wish the setpoint $\mathbf{w} = (w_1, w_2)^T$ to be such that w_1 corresponds to x_3 and w_2 corresponds to 3.28 x_1 (x_1 is, for example, a position expressed in meters and that we want w_2 to be expressed in feet, knowing that 1 meter = 3.28 feet). Calculate the setpoint matrix.

3) Calculate $\mathbf{H}$ in a way that $\mathbf{E}\mathbf{x}$ converges toward $\mathbf{x}_c$, again for a constant setpoint.

EXERCISE 4.12.– Control for a pump-operating motor

We consider the direct current motor of Figure 4.6.

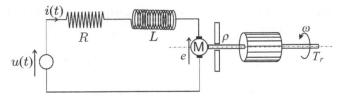

Figure 4.6. *Direct current motor*

Its state equations are of the form:

$$\begin{cases} \frac{di}{dt} = -\frac{R}{L}i - \frac{\kappa}{L}\omega + \frac{u}{L} \\ \dot{\omega} = \frac{\kappa}{J}i - \frac{\rho}{J}\omega - \frac{T_r}{J} \end{cases}$$

where κ, R, L, J are constant parameters of the motor. The inputs are the voltage u and the torque T_r, the state variables are i and ω. We use this motor to pump water from a well. In this case, the torque used is $T_r = \alpha\omega$, where α is a constant parameter. We will ignore ρ before α. There is consequently a single input for the motor+pump system: the voltage u.

1) Give the state equations of the motor+pump system.

2) We choose as output $y = \omega$. Calculate the transfer function of the motor+pump system.

3) Give the differential equation associated with this transfer function.

4) We will take as state vector $\mathbf{x} = (y, \dot{y})^T$. Give a state representation of the system in matrix form.

5) A *proportional-derivative* controller is a linear combination of the output y, its derivative $\dot{y}$ and the setpoint w (be careful not to confuse the setpoint w with the speed of the motor shaft ω). This controller is written in the form:

$$u = hw - k_1 y - k_2 \dot{y}$$

Give the state equations of the motor+pump system looped by such a controller.

6) Give the values of k_1 and k_2 that we need to choose in order to have the poles of the looped system equal to -1.

7) Find h in such a way that y converges toward w when the setpoint w is constant.

EXERCISE 4.13.– Proportional, integral and derivative control

We consider a second-order system of the form described by the following differential equation:

$$\ddot{y} + a_1 \dot{y} + a_0 y = u$$

1) Give the state equation of the system in matrix form. We will take as state vector $\mathbf{x} = (y \ \dot{y})^{\mathrm{T}}$.

2) Let w be a setpoint that we will assume constant. We would like $y(t)$ to converge toward w. We define the error by: $e(t) = w - y(t)$. We suggest controlling our system by the following proportional-derivative-integral (PID) controller:

$$u(t) = \alpha_{-1} \int_0^t e(\tau) \, d\tau \ + \ \alpha_0 e(t) \ + \ \alpha_1 \dot{e}(t)$$

where the α_i are the coefficients of the controller. This is a state feedback controller, where we assume that $\mathbf{x}(t)$ is measured. Give the state equations of this PID controller with the inputs $\mathbf{x}$, w and output u. A state variable $z(t) = \int_0^t e(\tau) \, d\tau$ will have to be created in order to take into account the integrator of the control.

3) Draw the wiring diagram of the looped system. This diagram will only be composed of integrators, adders and amplifiers. Encircle the controller on the one hand, and the system to be controlled on the other hand.

4) Give the state equations of the looped system in matrix form.

5) How do we choose the coefficients α_i of the control (as a function of the a_i) so as to have a stable looped system in which all the poles are equal to -1?

6) We slightly change the value of the parameters a_0 and a_1 while keeping the same controller. We assume that this modification does not destabilize our system. The new values for a_0, a_1 are denoted by a'_0, a'_1. For a value $\bar{w}$ for given w, what value $\bar{y}$ does converge y to? What do you conclude from this?

EXERCISE 4.14.– State feedback for an order 3 system

We consider the system described in Figure 4.7.

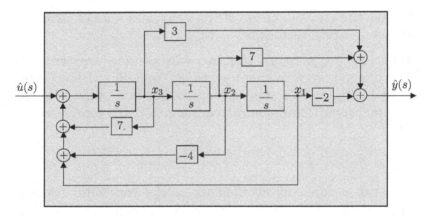

Figure 4.7. *State feedback for a linear system in canonical control form*

1) Give the state equations of this system. What is its characteristic polynomial?

2) We would like to control this system by a state feedback of the form $u = -\mathbf{K}\mathbf{x} + hw$, where w is the setpoint. Calculate $\mathbf{K}$ in order for all the poles to be equal to -1.

3) We would like, at equilibrium (in other words, when the setpoint and the output no longer change), to have $y = w$. Deduce from this the value to take for the precompensator h.

EXERCISE 4.15.– State feedback with integral effect, monovariate case

We consider the system described by the state equations:

$$\begin{cases} \dot{\mathbf{x}} = \begin{pmatrix} 1 & 1 \\ 0 & 2 \end{pmatrix} \mathbf{x} + \begin{pmatrix} 0 \\ 1 \end{pmatrix} u \\ y = x_1 \end{cases}$$

where u is the input, y is the output and x is the state vector.

1) Give the characteristic polynomial of the system. Is the system stable?

2) We loop the system by the following state feedback control:

$$u = \alpha \int_0^t (w(\tau) - x_1(\tau))\, d\tau - \mathbf{Kx}, \text{ with } \mathbf{K} = (k_1 \; k_2)$$

where w is the setpoint. Give the state equations of the controller (we will denote by z the state variable of the controller). What are the poles of the controller?

3) Give the state equations of the looped system.

4) Calculate **K** and α in order for all the poles to be equal to -1.

5) We choose a setpoint $w = \bar{w}$ constant in time. What values $\bar{\mathbf{x}}$ and $\bar{z}$ does the state of the system x and state of the controller z tend to? What value $\bar{y}$ does the output y tend to?

6) We now replace the evolution matrix **A** by another matrix $\bar{\mathbf{A}}$ close to **A**, while keeping the same controller. What value y will it converge to?

EXERCISE 4.16.– State feedback with integral effect, general case

We consider the system described by the state equation $\dot{\mathbf{x}} = \mathbf{Ax} + \mathbf{Bu} + \mathbf{p}$ where **p** is an unknown and constant disturbance vector that represents an external disturbance. We will take $m = \dim \mathbf{u}$ and $n = \dim \mathbf{x}$. A state feedback controller with integral effect is of the form:

$$\mathbf{u} = \mathbf{K}_i \int_0^t (\mathbf{w} - \mathbf{x}_c)\, dt - \mathbf{K}\mathbf{x}$$

where w is the setpoint and $\mathbf{x}_c$ is a vector with same dimension as u representing the setpoint state variables (in other words, those we wish to control directly by using w). The vector $\mathbf{x}_c$ is linked to the state x by the relation $\mathbf{x}_c = \mathbf{E}\mathbf{x}$, where E is a known matrix.

1) Give the state equations of the looped system. Recall that the state of the looped system will be composed of the state x of the system to control and of the vector z of the values memorized in the integrators. Give the dimensions of the values w, $\mathbf{x}_c$, p, z, A, B, E, $\mathbf{K}_i$, K.

2) Show, with the MATLAB `place` function, how we can choose the matrices K and $\mathbf{K}_i$ so that all the poles of the looped system are equal to -1.

3) Show that, for a constant setpoint w and disturbance p, we necessarily have $\mathbf{x}_c = \mathbf{w}$, once the steady state has been reached. What is then the value of p (as a function of x and z)?

4) Give the state equations of the controller.

EXERCISE 4.17.– Observer

We consider the system with input u and output y of Figure 4.8.

1) Give, in matrix form, a state representation for this system. Deduce from this a simplification for the wiring system.

2) Study the stability of this system.

3) Give the transfer function of this system.

4) Is the system controllable? Is it observable?

5) Find an observer that allows us to generate a state $\hat{x}(t)$, such that the error $||\hat{x}-x||$ converges toward zero at e^{-t} (which means to place all the poles at -1). Give this observer in state equation form.

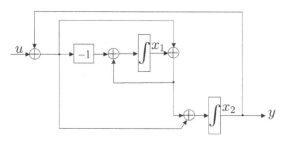

Figure 4.8. *Second-order system for which we need to design an observer*

EXERCISE 4.18.– Amplitude demodulator

A sinusoidal signal with pulsation ω can be written in the form:

$$y(t) = a\cos(\omega t + b)$$

where the parameters a and b are unknown. From the measure of $y(t)$, we need to find the amplitude a of the signal $y(t)$.

1) Find an order 2 state equation capable of generating the signal $y(t)$. We will take as state variables $x_1 = \omega y$ and $x_2 = \dot{y}$.

2) Let us assume that at time t, we know the state vector $\mathbf{x}(t)$. Deduce from this an expression of the amplitude a of the signal $y(t)$ as a function of $\mathbf{x}(t)$.

3) We only measure $y(t)$. Propose a state observer (by a pole placement method) that generates us an estimation $\hat{\mathbf{x}}(t)$ of the state $\mathbf{x}(t)$. We will place all the poles at -1.

4) Deduce from this the state equations of an estimator with input y and output $\hat{a}$ that gives us an estimation $\hat{a}$ of the amplitude a of a sinusoidal signal with pulsation ω.

5) In MATLAB, generate a sampled signal:

$$y(t) = a\cos\omega t + n(t)$$

with $t \in \{0, dt, 2dt, \ldots, 10\}$ where $dt = 0.01$ is the sampling period. The signal $n(t)$ is a Gaussian noise with standard deviation 0.1. This noise is assumed white, in other words that $n(t_1)$ is independent of $n(t_2)$ for $t_1 \neq t_2$. We will take an amplitude $a = 2$ and a pulsation $\omega = 1$. By using the state observer developed in previous exercises, create a filter which, from $y(t)$, estimates the amplitude a of the signal. Draw the filtered signal $\hat{y}(t)$ as well as the estimation $\hat{a}(t)$ of the amplitude.

6) Draw in MATLAB the Bode plot of the filter that generates $\hat{y}(t)$ from $y(t)$. Discuss.

EXERCISE 4.19.– Output feedback of a non-strictly proper system

We consider the system $\mathcal{S}$ with input u and output y described by the differential equation:

$$\dot{y} - 2y = \dot{u} + 3u$$

Since in this case the degree of differentiation of u is not strictly smaller than that of y, the system is not strictly proper.

1) Write this system in a state representation form.

2) In order to make this system strictly proper (in other words, with a direct matrix D equal to zero), we create a new output $z = y - \alpha u$. Give the appropriate value of α. We will denote by $\mathcal{S}_z$ this new system whose input is u and output is z.

3) Find an output feedback controller for $\mathcal{S}_z$ that sets all the poles of the looped system to -1.

4) Deduce from this a controller $\mathcal{S}_r$ (in state equation form) for $\mathcal{S}$ that sets all the poles of the looped system to -1. We will denote by $\mathcal{S}_b$ the system $\mathcal{S}$ looped by the controller $\mathcal{S}_r$.

5) Give the state equations of the looped system $\mathcal{S}_b$.

6) Calculate the transfer function of the looped system $\mathcal{S}_b$.

7) Calculate the static gain of the looped system. This gain corresponds to the ratio $\frac{\bar{y}}{\bar{w}}$ in a steady state. Deduce from this the gain of the precompensator to be placed before the system that would allow us to have a static gain of 1.

EXERCISE 4.20.– Output feedback with integral effect

We consider the system described by the state equation:

$$\begin{cases} \dot{\mathbf{x}} = \mathbf{A}\mathbf{x} + \mathbf{B}\mathbf{u} + \mathbf{p} \\ \mathbf{y} = \mathbf{C}\mathbf{x} \end{cases}$$

where p is a disturbance vector, representing an external disturbance that could not be taken into account in the modeling (wind, the slope of the terrain, the weight of the people in an elevator, etc.). The vector p is assumed to be known and constant. We will take $m = \dim \mathbf{u}$, $n = \dim \mathbf{x}$ and $p = \dim \mathbf{y}$. We need to control this system using the output feedback controller with integral effect represented in Figure 4.9.

In this figure, w is the setpoint and $\mathbf{y}_c$ is a vector with the same dimension as u representing the setpoint output variables (in other words, those that we wish to control directly using w and thus have $\mathbf{y}_c = \mathbf{w}$, when the setpoint is constant). The vector $\mathbf{y}_c$ satisfies the relation $\mathbf{y}_c = \mathbf{E}\mathbf{y}$, where E is a known matrix.

1) Give the state equations of the looped system in matrix form.

2) Let us take $\varepsilon = \hat{\mathbf{x}} - \mathbf{x}$. Express these state equations in matrix form by taking, this time, the state vectors (x, z, ε).

3) By using the MATLAB place function, show how we can arbitrarily set all the poles of the looped system.

4) Show that, for a constant setpoint w and disturbance p, in a steady state, we necessarily have $y_c = w$.

5) Give the state equations of the controller, in matrix form.

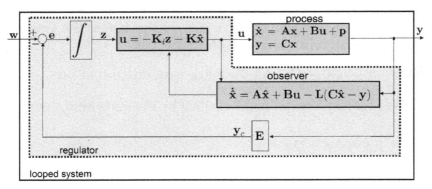

Figure 4.9. *Output feedback controller with integral effect*

4.6. Solutions

Solution to Exercise 4.1 (removing the direct matrix)

The new system has the state equations:

$$\begin{cases} \dot{x} = Ax + Bu \\ z = Cx \end{cases}$$

The direct matrix D has, therefore, disappeared.

Solution to Exercise 4.2 (non-observable states and non-controllable states)

1) The transfer function of the system is given by:

$$\begin{pmatrix} 1 & 1 & 0 & 0 \end{pmatrix} \left(\begin{pmatrix} s & 0 & 0 & 0 \\ 0 & s & 0 & 0 \\ 0 & 0 & s & 0 \\ 0 & 0 & 0 & s \end{pmatrix} - \begin{pmatrix} -1 & 1 & 0 & 0 \\ 0 & 1 & 0 & 0 \\ 1 & 1 & 1 & 1 \\ 0 & 1 & 0 & 1 \end{pmatrix} \right)^{-1} \begin{pmatrix} 1 \\ 0 \\ 1 \\ 0 \end{pmatrix} + 1 = \frac{s+2}{s+1}$$

Of course, the inversion is not necessary and the following, faster calculation can be performed:

$$\begin{cases} \hat{y} = \hat{x}_1 + \hat{x}_2 + \hat{u} \\ s\hat{x}_1 = -\hat{x}_1 + \hat{x}_2 + \hat{u} \\ s\hat{x}_2 = \hat{x}_2 \end{cases} \Leftrightarrow \begin{cases} \hat{y} = \hat{x}_1 + \hat{u} \\ s\hat{x}_1 = -\hat{x}_1 + \hat{u} \\ \hat{x}_2 = 0 \end{cases} \Rightarrow \hat{y} = \left(\frac{1}{s+1} + 1\right)\hat{u}$$

2) The characteristic polynomial of the evolution matrix is $P(s) = s^4 - 2s^3 + 2s - 1$. Its roots are $\{-1, 1, 1, 1\}$. The system is therefore unstable, even though its transfer function $\frac{s+2}{s+1}$ is stable.

3) The wiring system (not represented here) shows that x_1 is observable and controllable, x_2 is observable and non-controllable, x_3 is non-observable and controllable, x_4 is non-observable and non-controllable.

4) The poles of the system are $1, 1, 1, -1$. The transmission pole is -1. The hidden modes are $-1, -1, -1$.

Solution to Exercise 4.3 (using the controllability and observability criteria)

1) The controllability matrix of the system is given by:

$$\mathbf{\Gamma}_{\text{con}} = \left(\mathbf{B} \mid \mathbf{AB} \mid \mathbf{A}^2\mathbf{B}\right) = \begin{pmatrix} 0 & 0 & 1 & 0 & 2 & 0 \\ 1 & 0 & 1 & 0 & 1 & 0 \\ 1 & a & 1 & a & 1 & a \end{pmatrix}$$

If $a = 0$, the rank is equal to 2, and therefore the system is non-controllable. For $a \neq 0$, $\text{rank}(\mathbf{\Gamma}_{\text{con}}) = 3$ and therefore the system is controllable.

2) The observability matrix is:

$$\Gamma_{obs} = \begin{pmatrix} C \\ CA \\ CA^2 \end{pmatrix} = \begin{pmatrix} 1 & 1 & b \\ 0 & 1 & 0 \\ 1 & 2 & b \\ 0 & 1 & 0 \\ 1 & 3 & b \\ 0 & 1 & 0 \end{pmatrix}$$

Its rank is equal to 2, for every value of b. The system is, therefore, not observable. We would have needed rank(Γ_{obs}) = 3 in order to have an observable system.

Solution to Exercise 4.4 (proof of the controllability criterion in discrete time)

REMINDER.– Let us consider the linear system $Ax = b$, where $x \in \mathbb{R}^n$ and the matrix A is horizontal-rectangular (with its number of columns n higher than its number of rows m). If A is of full rank (in other words, equal to m) then, the set of solutions is an affine space of dimension $n - m$.

1) Let us note first of all that if $x(n)$ can be chosen as desired, from any initial vector, it seems to be clear that it is possible to move in the state space, and therefore that the system is controllable. We have:

$x(1) = Ax(0) + Bu(0)$

$x(2) = Ax(1) + Bu(1) = A^2x(0) + ABu(0) + Bu(1)$

$x(3) = Ax(2) + Bu(2) = A^3x(0) + A^2Bu(0) + ABu(1) + Bu(2)$

$\vdots$

$x(n) = A^n x(0) + A^{n-1}Bu(0) + A^{n-2}Bu(1) + \ldots$
$\quad + ABu(n-2) + Bu(n-1).$

Thus:

$$\mathbf{x}(n) = \mathbf{A}^n \mathbf{x}(0) + \underbrace{(\mathbf{B} \mid \mathbf{AB} \mid \ldots \mid \mathbf{A}^{n-1}\mathbf{B})}_{\Gamma_{\text{con}}} \begin{pmatrix} \mathbf{u}(n-1) \\ \mathbf{u}(n-2) \\ \vdots \\ \mathbf{u}(1) \\ \mathbf{u}(0) \end{pmatrix}$$

$$= \mathbf{A}^n \mathbf{x}(0) + \Gamma_{\text{con}} \mathbf{v}$$

where Γ_{con} is the controllability matrix of dimension $n \times mp$ and v is the vector of dimension mp of inputs (and future inputs) that we are looking for.

2) In order to impose $\mathbf{x}(n)$ arbitrarily, we need to solve the system of n linear equations:

$$\Gamma_{\text{con}} \mathbf{v} = \mathbf{x}(n) - \mathbf{A}^n \mathbf{x}(0)$$

with mp unknowns. If Γ_{con} is of full rank (in other words, that its column vectors form a generating family of $\mathbb{R}^n$), then there is always at least one solution v for this system.

Solution to Exercise 4.5 (proof of the controllability criterion in continuous time)

1) We have:

$$\mathbf{z}^\mathrm{T}.\mathbf{x}(t) = \mathbf{z}^\mathrm{T} e^{\mathbf{A}t}\mathbf{x}(0) + \int_0^t \mathbf{z}^\mathrm{T} e^{\mathbf{A}(t-\tau)} \mathbf{B}\mathbf{u}(\tau) d\tau$$

where z is a non-zero vector such that $\mathbf{z}^\mathrm{T}.\Gamma_{\text{con}} = 0$. It is therefore sufficient to show that the quantity $\mathbf{z}^\mathrm{T} e^{\mathbf{A}(t-\tau)} \mathbf{B}$ is always zero. We have:

$$e^{\mathbf{A}(t-\tau)} = \sum_{i=0}^{\infty} \frac{(t-\tau)^i \mathbf{A}^i}{i!}$$

However, the Cayley–Hamilton theorem tells us that the matrix **A** satisfies the characteristic equation:

$$\mathbf{A}^n + a_{n-1}\mathbf{A}^{n-1} + \cdots + a_2\mathbf{A}^2 + a_1\mathbf{A} + a_0\mathbf{I} = \mathbf{0}$$

where $s^n + a_{n-1}s^{n-1} + \cdots + a_1 s + a_0$ is the characteristic polynomial of **A**. From this, we deduce that all the powers of **A** can be expressed as linear combinations of $\mathbf{A}^i, i \in \{0, \ldots, n-1\}$. Thus, $e^{\mathbf{A}(t-\tau)}$ can be written in the form:

$$e^{\mathbf{A}(t-\tau)} = \sum_{i=0}^{n-1} \beta_i \cdot (t-\tau)^i \mathbf{A}^i$$

We, therefore, have:

$$\mathbf{z}^T e^{\mathbf{A}(t-\tau)} \mathbf{B} = \mathbf{z}^T \left(\sum_{i=0}^{n-1} \beta_i \cdot (t-\tau)^i \mathbf{A}^i \right) \cdot \mathbf{B}$$

$$= \sum_{i=0}^{n-1} \beta_i \cdot (t-\tau)^i \cdot \left(\mathbf{z}^T \mathbf{A}^i \mathbf{B} \right)$$

Since the quantity $\mathbf{z}^T \mathbf{A}^i \mathbf{B}$ is equal to zero (by assumption), $\mathbf{z}^T e^{\mathbf{A}(t-\tau)} \mathbf{B} = \mathbf{0}$. Thus, for all **u** we have $\mathbf{z}^T . \mathbf{x}(t) = \mathbf{z}^T e^{\mathbf{A}t} \mathbf{x}(0)$, and therefore the control **u** cannot influence the component of **x** following **z**. The system is, therefore, non-controllable.

2) We will now show that if $\text{rank}(\Gamma_{\text{con}}) = n$, for every pair $(\mathbf{x}(0), \mathbf{x}(1))$, there exists a polynomial control $\mathbf{u}(t), t \in [0,1]$ of the form:

$$\mathbf{u}(t) = \mathbf{m}(0) + \mathbf{m}(1)t + \cdots + \mathbf{m}(n-1)t^{n-1}$$

$$= \underbrace{\begin{pmatrix} \mathbf{m}(0) & \mathbf{m}(1) & \cdots & \mathbf{m}(n-1) \end{pmatrix}}_{\mathbf{M}} \begin{pmatrix} 1 \\ t \\ \vdots \\ t^{n-1} \end{pmatrix}$$

which leads the system from the state x(0) to the state x(1). We have:

$$x(1) = e^{\mathbf{A}}x(0) + \int_0^1 e^{\mathbf{A}(1-\tau)}\mathbf{B}u(\tau)d\tau$$

$$= e^{\mathbf{A}}\left(x(0) + \int_0^1 e^{-\tau\mathbf{A}}\mathbf{B}u(\tau)d\tau\right)$$

We, therefore, need to find u such that:

$$\int_0^1 e^{-\tau\mathbf{A}}\mathbf{B}u(\tau)d\tau = \underbrace{e^{-\mathbf{A}}x(1) - x(0)}_{\mathbf{z}}$$

However, $e^{-\tau\mathbf{A}}$ can be written in the form:

$$e^{-\tau\mathbf{A}} = \sum_{i=0}^{n-1} \gamma_i.\tau^i.\mathbf{A}^i$$

(we need to apply the Cayley–Hamilton theorem on the matrix $\tau\mathbf{A}$ this time). Thus:

$$\int_0^1 e^{-\tau\mathbf{A}}\mathbf{B}u(\tau)d\tau = \int_0^1 \sum_{i=0}^{n-1} \gamma_i.\tau^i.\mathbf{A}^i.\mathbf{B}.u(\tau)d\tau$$

$$= \sum_{i=0}^{n-1} \mathbf{A}^i.\mathbf{B}.\gamma_i \underbrace{\int_0^1 \tau^i.u(\tau)d\tau}_{\mathbf{h}(i)}$$

Since the controllability matrix is of rank n and that the γ_i are non-zero (which we will assume), the system to be solved:

$$\sum_{i=0}^{n-1} \mathbf{A}^i.\mathbf{B}.\gamma_i.\mathbf{h}(i) = \mathbf{z}$$

always has a solution $\mathbf{h}(0), \ldots, \mathbf{h}(n-1)$. We, therefore, need to solve the following n equations:

$$\int_0^1 \tau^i . \mathbf{u}(\tau) d\tau = \mathbf{h}(i), \ i = 0, \ldots, n-1$$

However:

$$\begin{aligned}
\int_0^1 \tau^i . \mathbf{u}(\tau) d\tau &= \int_0^1 \tau^i . \left(\sum_{k=0}^{n-1} \mathbf{m}(k) . \tau^k \right) d\tau \\
&= \sum_{k=0}^{n-1} \mathbf{m}(k) \int_0^1 \tau^{i+k} d\tau \\
&= \sum_{k=0}^{n-1} \frac{1}{k+i+1} \mathbf{m}(k) \\
&= \begin{pmatrix} \mathbf{m}(0) & \mathbf{m}(1) & & \mathbf{m}(n-1) \end{pmatrix} \begin{pmatrix} \frac{1}{i+1} \\ \frac{1}{i+2} \\ \\ \frac{1}{i+n} \end{pmatrix}
\end{aligned}$$

Thus, the n equations to be solved are written as:

$$\underbrace{\begin{pmatrix} \mathbf{m}(0) & & \mathbf{m}(n-1) \end{pmatrix}}_{\mathbf{M}} \cdot \underbrace{\begin{pmatrix} 1 & \frac{1}{2} & & \frac{1}{n} \\ \frac{1}{2} & \frac{1}{3} & & \frac{1}{n+1} \\ & & & \\ \frac{1}{n} & \frac{1}{n+1} & & \frac{1}{2n-1} \end{pmatrix}}_{\mathbf{T}} = \underbrace{\begin{pmatrix} \mathbf{h}(0) & & \mathbf{h}(n-1) \end{pmatrix}}_{\mathbf{H}}$$

The matrix $\mathbf{T}$ is invertible, and therefore $\mathbf{M} = \mathbf{HT}^{-1}$, which gives us the control u.

Solution to Exercise 4.6 (proof of the observability criterion)

In this exercise, we will show that, in the case that our system is observable, the knowledge of the first $n-1$ derivatives of the outputs and the $n-2$ derivatives of the inputs allows us to find the state vector.

1) Let us differentiate the observation equation $n-1$ times. We obtain:

$$y = Cx$$
$$\dot{y} = CAx + CBu$$
$$\ddot{y} = CA^2x + CABu + CB\dot{u}$$
$$\vdots$$
$$y^{(n-1)} = CA^{n-1}x + CA^{n-2}Bu + CA^{n-3}B\dot{u} + \ldots$$
$$+ CABu^{(n-3)} + CBu^{(n-2)}.$$

Or, in matrix form:

$$\begin{pmatrix} y \\ \dot{y} \\ \ddot{y} \\ \vdots \\ y^{(n-1)} \end{pmatrix} = \begin{pmatrix} C \\ CA \\ CA^2 \\ \vdots \\ CA^{n-1} \end{pmatrix} x + \begin{pmatrix} 0 & 0 & 0 & 0 \\ CB & 0 & 0 & 0 \\ CAB & CB & 0 & \\ \vdots & & \ddots & \ddots \\ CA^{n-2}B & \cdots & CAB & CB \end{pmatrix} \begin{pmatrix} u \\ \dot{u} \\ \ddot{u} \\ \vdots \\ u^{(n-2)} \end{pmatrix}$$

2) This equation can be written in the form $z = \Gamma_{obs}x + \Phi v$, where z is the vector of all the outputs and their derivatives, v is the vector of all the inputs and their derivatives, Γ_{obs} is the observability matrix and Φ is the remaining matrix. The system we need to solve in order to reach state x is given by:

$$\Gamma_{obs}x = z - \Phi v$$

This equation has at most one solution if Γ_{obs} is of full rank. The absence of a solution would mean that v and z are incompatible with the equations of our system, which is incompatible with our assumptions. This solution is given by:

$$x = \left(\Gamma_{obs}^T \Gamma_{obs}\right)^{-1} \Gamma_{obs}^T \cdot (z - \Phi v)$$

The matrix $\left(\Gamma_{\text{obs}}^T \Gamma_{\text{obs}}\right)^{-1} \Gamma_{\text{obs}}^T$ is called the pseudoinverse of the matrix Γ_{obs}. It only exists if Γ_{obs} is of full rank.

Solution to Exercise 4.7 (Kalman decomposition)

The requested diagram is represented in Figure 4.10. We can see the four subsystems: $\mathcal{S}_1$ (controllable and observable), $\mathcal{S}_2$ (non-controllable and observable), $\mathcal{S}_3$ (controllable and non-observable) and $\mathcal{S}_4$ (neither controllable nor observable).

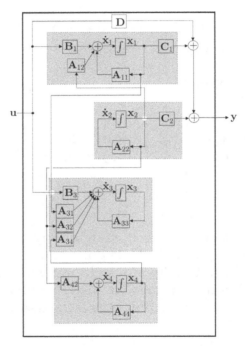

Figure 4.10. *Details of the Kalman decomposition for the linear systems*

Solution to Exercise 4.8 (resolution of the pole placement equation)

We have:

$$P_{\text{con}}(s) = (s+1)(s+1) = s^2 + 2s + 1$$

The pole placement equation $\det(s\mathbf{I} - \mathbf{A} + \mathbf{BK}) = P_{\text{con}}(s)$ is written as:

$$\det\left(\begin{pmatrix} s & 0 \\ 0 & s \end{pmatrix} - \begin{pmatrix} 1 & 2 \\ 3 & 4 \end{pmatrix} + \begin{pmatrix} 1 \\ 2 \end{pmatrix}\begin{pmatrix} k_1 & k_2 \end{pmatrix}\right) = s^2 + 2s + 1$$

i.e.:

$$\det\begin{pmatrix} s + k_1 - 1 & k_2 - 2 \\ 2k_1 - 3 & s + 2k_2 - 4 \end{pmatrix} = s^2 + 2s + 1$$

or, alternatively:

$$s^2 + s(k_1 + 2k_2 - 5) + k_2 - 2 = s^2 + 2s + 1$$

We obtain the following linear system:

$$\begin{pmatrix} 1 & 2 \\ 0 & 1 \end{pmatrix}\begin{pmatrix} k_1 \\ k_2 \end{pmatrix} + \begin{pmatrix} -5 \\ -2 \end{pmatrix} = \begin{pmatrix} 2 \\ 1 \end{pmatrix}$$

Let $\mathbf{K} = (k_1 \ k_2) = (1 \ 3)$. We could have obtained this result directly by using the `ppol` instruction of SCILAB or `place` of MATLAB.

Solution to Exercise 4.9 (output feedback of a scalar system)

1) In order to find K and L, we need to solve:

$$\det(s - 3 + 2K) = s + 1$$
$$\det(s - 3 + 4L) = s + 1$$

We obtain $K = 2$ and $L = 1$. For the calculation of the precompensator, we will take $E = 1$ (since the setpoint variable is $x_c = x$). Thus:

$$H = -\left(E(A - BK)^{-1}B\right)^{-1} = \frac{-1}{1 \cdot (3 - 2 \cdot 2)^{-1} \cdot 2} = \frac{1}{2}$$

The controller we are looking for is, therefore, given by:

$$\mathcal{R} := \begin{cases} \frac{d}{dt}\hat{x} = -5\hat{x} + w + y \\ u = -2\hat{x} + \frac{1}{2}w \end{cases}$$

2) The looped system is described by the following evolution equations:

$$\begin{cases} \dot{x} = 3x - 4\hat{x} + w \\ \frac{d}{dt}\hat{x} = 4x - 5\hat{x} + w \end{cases}$$

We can then verify whether the poles of this system are the ones we have placed (more precisely -1 and -1), which is a consequence of the separation principle (see Exercise 4.10).

Solution to Exercise 4.10 (separation principle)

1) The controller has as state vector $\hat{\mathbf{x}}$ and as input the setpoint w and the output of the system to control y. The output of the controller is the control u. The state representation of the controller is expressed by:

$$\begin{cases} \frac{d}{dt}\hat{\mathbf{x}} = \mathbf{A}\hat{\mathbf{x}} + \mathbf{B}(\mathbf{H}\mathbf{w} - \mathbf{K}\hat{\mathbf{x}}) - \mathbf{L}(\mathbf{C}\hat{\mathbf{x}} - \mathbf{y}) \\ \mathbf{u} = \mathbf{H}\mathbf{w} - \mathbf{K}\hat{\mathbf{x}} \end{cases}$$

or, in a simplified form:

$$\begin{cases} \frac{d}{dt}\hat{\mathbf{x}} = (\mathbf{A} - \mathbf{B}\mathbf{K} - \mathbf{L}\mathbf{C})\hat{\mathbf{x}} + \mathbf{B}\mathbf{H}\mathbf{w} + \mathbf{L}\mathbf{y} \\ \mathbf{u} = -\mathbf{K}\hat{\mathbf{x}} + \mathbf{H}\mathbf{w} \end{cases}$$

In matrix form, this equation is written as:

$$\begin{cases} \frac{d}{dt}\hat{\mathbf{x}} = (\mathbf{A} - \mathbf{B}\mathbf{K} - \mathbf{L}\mathbf{C})\hat{\mathbf{x}} + \begin{pmatrix} \mathbf{B}\mathbf{H} & \mathbf{L} \end{pmatrix} \begin{pmatrix} \mathbf{w} \\ \mathbf{y} \end{pmatrix} \\ \mathbf{u} = -\mathbf{K}\hat{\mathbf{x}} + \begin{pmatrix} \mathbf{H} & \mathbf{0} \end{pmatrix} \begin{pmatrix} \mathbf{w} \\ \mathbf{y} \end{pmatrix} \end{cases}$$

These are the equations we need to wire or program in order to control our system.

2) The state equations associated with the looped system with input w and output y of our output feedback controlled system are given by:

$$\begin{cases} \dot{x} = Ax + B(w - K\hat{x}) \\ \frac{d\hat{x}}{dt} = (A - BK - LC)\hat{x} + LCx + Bw \\ y = Cx \end{cases}$$

In matrix form, these equations are written as:

$$\begin{cases} \frac{d}{dt}\begin{pmatrix} x \\ \hat{x} \end{pmatrix} = \begin{pmatrix} A & -BK \\ LC & A - BK - LC \end{pmatrix}\begin{pmatrix} x \\ \hat{x} \end{pmatrix} + \begin{pmatrix} B \\ B \end{pmatrix} w \\ y = \begin{pmatrix} C & 0 \end{pmatrix}\begin{pmatrix} x \\ \hat{x} \end{pmatrix} \end{cases}$$

3) Let us take $\varepsilon_x = \hat{x} - x$. Since:

$$\begin{pmatrix} x \\ \varepsilon_x \end{pmatrix} = \begin{pmatrix} I & 0 \\ -I & I \end{pmatrix}\begin{pmatrix} x \\ \hat{x} \end{pmatrix}$$

or equivalently:

$$\begin{pmatrix} x \\ \hat{x} \end{pmatrix} = \begin{pmatrix} I & 0 \\ I & I \end{pmatrix}\begin{pmatrix} x \\ \varepsilon_x \end{pmatrix}$$

another possible state vector for the looped system is $(\mathbf{x}, \varepsilon_x)$. After change of basis, the state equations become:

$$\begin{cases} \begin{pmatrix} \dot{\mathbf{x}} \\ \dot{\varepsilon}_x \end{pmatrix} = \begin{pmatrix} \mathbf{I} & 0 \\ -\mathbf{I} & \mathbf{I} \end{pmatrix} \begin{pmatrix} \mathbf{A} & -\mathbf{BK} \\ \mathbf{LC} & \mathbf{A}-\mathbf{BK}-\mathbf{LC} \end{pmatrix} \begin{pmatrix} \mathbf{I} & 0 \\ \mathbf{I} & \mathbf{I} \end{pmatrix} \begin{pmatrix} \mathbf{x} \\ \varepsilon_x \end{pmatrix} \\ \qquad + \begin{pmatrix} \mathbf{I} & 0 \\ -\mathbf{I} & \mathbf{I} \end{pmatrix} \begin{pmatrix} \mathbf{B} \\ \mathbf{B} \end{pmatrix} \mathbf{w} \\ \mathbf{y} \;= \begin{pmatrix} \mathbf{C} & 0 \end{pmatrix} \begin{pmatrix} \mathbf{I} & 0 \\ \mathbf{I} & \mathbf{I} \end{pmatrix} \begin{pmatrix} \mathbf{x} \\ \varepsilon_x \end{pmatrix} \end{cases}$$

or, alternatively:

$$\begin{cases} \begin{pmatrix} \dot{\mathbf{x}} \\ \dot{\varepsilon}_x \end{pmatrix} = \begin{pmatrix} \mathbf{A}-\mathbf{BK} & -\mathbf{BK} \\ 0 & \mathbf{A}-\mathbf{LC} \end{pmatrix} \begin{pmatrix} \mathbf{x} \\ \varepsilon_x \end{pmatrix} + \begin{pmatrix} \mathbf{B} \\ 0 \end{pmatrix} \mathbf{w} \\ \mathbf{y} = \begin{pmatrix} \mathbf{C} & 0 \end{pmatrix} \begin{pmatrix} \mathbf{x} \\ \varepsilon_x \end{pmatrix} \end{cases}$$

Let us note that the input w cannot act on ε_x, which is compatible with the fact that ε_x is not a controllable subvector.

4) The characteristic polynomial of the looped system is:

$$P(s) = \det\left(s\mathbf{I} - \begin{pmatrix} \mathbf{A}-\mathbf{BK} & \mathbf{BK} \\ 0 & \mathbf{A}-\mathbf{LC} \end{pmatrix} \right)$$

$$= \det \begin{pmatrix} s\mathbf{I}-\mathbf{A}+\mathbf{BK} & -\mathbf{BK} \\ 0 & s\mathbf{I}-\mathbf{A}+\mathbf{LC} \end{pmatrix}$$

$$= \det(s\mathbf{I}-\mathbf{A}+\mathbf{BK}) \cdot \det(s\mathbf{I}-\mathbf{A}+\mathbf{LC}) = P_{\text{con}}(s) \cdot P_{\text{obs}}(s).$$

Therefore, the poles of the looped system are composed of the poles placed for the control and the poles placed for the observation. This is the *separation principle*.

Solution to Exercise 4.11 (choosing the precompensator)

1) We have:

$$0 = \begin{pmatrix} A - BK & -BK \\ 0 & A - LC \end{pmatrix} \begin{pmatrix} \bar{x} \\ \bar{\varepsilon}_x \end{pmatrix} + \begin{pmatrix} BH \\ 0 \end{pmatrix} \bar{w}$$

Since $(A - LC)$ is invertible (since all the poles placed for the observer are strictly stable), $\bar{\varepsilon}_x$ is necessarily nil. The previous equation becomes:

$$(A - BK)\bar{x} + BH\bar{w} = 0$$

Since $A - BK$ is also invertible (since all the poles placed for the control are strictly stable), we have:

$$\bar{x} = -(A - BK)^{-1} BH\bar{w}$$

2) Since $x_{c1} = x_3$ and $x_{c2} = 3.28 x_1$, we need to take:

$$E = \begin{pmatrix} 0 & 0 & 1 \\ 3.28 & 0 & 0 \end{pmatrix}$$

3) At equilibrium, we have:

$$\bar{x}_c = E\bar{x} = -E(A - BK)^{-1} BH\bar{w}$$

Thus:

$$\bar{x}_c = \bar{w} \Leftrightarrow -E(A - BK)^{-1} BH = I \Leftrightarrow H$$
$$= -\left(E(A - BK)^{-1} B\right)^{-1}$$

The insertion of a precompensator, therefore, allows us to assign, to each setpoint composing w, a particular state variable. The variables set in this manner can then be controlled independently of one another.

Solution to Exercise 4.12 (control for a pump-operating motor)

1) We have:

$$\begin{cases} \frac{di}{dt} = -\frac{R}{L}i - \frac{\kappa}{L}\omega + \frac{u}{L} \\ \dot{\omega} = \frac{\kappa}{J}i - \frac{\rho+\alpha}{J} \end{cases}$$

Since ρ is negligible compare to α, these equations can be written in the following matrix form:

$$\frac{d}{dt}\begin{pmatrix} i \\ \omega \end{pmatrix} = \begin{pmatrix} -\frac{R}{L} & -\frac{\kappa}{L} \\ \frac{\kappa}{J} & -\frac{\alpha}{J} \end{pmatrix} \begin{pmatrix} i \\ \omega \end{pmatrix} + \begin{pmatrix} \frac{1}{L} \\ 0 \end{pmatrix} u$$

$$y = \begin{pmatrix} 0 & 1 \end{pmatrix} \begin{pmatrix} i \\ \omega \end{pmatrix}$$

2) The transfer function is:

$$G(s) = C\left(sI - A\right)^{-1} B + D$$

$$= \begin{pmatrix} 0 & 1 \end{pmatrix} \left(s \begin{pmatrix} 1 & 0 \\ 0 & 1 \end{pmatrix} - \begin{pmatrix} -\frac{R}{L} & -\frac{\kappa}{L} \\ \frac{\kappa}{J} & -\frac{\alpha}{J} \end{pmatrix} \right)^{-1} \begin{pmatrix} \frac{1}{L} \\ 0 \end{pmatrix}$$

$$= \frac{\kappa}{JLs^2 + (L\alpha + JR)s + R\alpha + \kappa^2} = \frac{\frac{\kappa}{JL}}{s^2 + \frac{L\alpha+JR}{JL}s + \frac{R\alpha+\kappa^2}{JL}}$$

3) The differential equation associated with this transfer function is:

$$\ddot{y} + \frac{L\alpha + JR}{JL}\dot{y} + \frac{R\alpha + \kappa^2}{JL}y = \frac{\kappa}{JL}u$$

4) By taking as state vector $\mathbf{x} = (y, \dot{y})^{\mathrm{T}}$, the state representation of the system is written as:

$$\begin{cases} \dot{\mathbf{x}} = \begin{pmatrix} 0 & 1 \\ -\frac{R\alpha+\kappa^2}{JL} & -\frac{L\alpha+JR}{JL} \end{pmatrix} \mathbf{x} + \begin{pmatrix} 0 \\ \frac{\kappa}{JL} \end{pmatrix} u \\ y = \begin{pmatrix} 1 & 0 \end{pmatrix} \mathbf{x} \end{cases}$$

5) We have:

$$\dot{\mathbf{x}} = \mathbf{A}\mathbf{x} + \mathbf{B}(hw - \mathbf{K}\mathbf{x}) = (\mathbf{A} - \mathbf{B}\mathbf{K})\mathbf{x} + \mathbf{B}hw$$

i.e.:

$$\begin{cases} \dot{\mathbf{x}} = \begin{pmatrix} 0 & 1 \\ -\frac{R\alpha+\kappa^2}{JL} - k_1\frac{\kappa}{JL} & -\frac{L\alpha+JR}{JL} - k_2\frac{\kappa}{JL} \end{pmatrix} \mathbf{x} + \begin{pmatrix} 0 \\ \frac{\kappa h}{JL} \end{pmatrix} w \\ y = \begin{pmatrix} 1 & 0 \end{pmatrix} \mathbf{x} \end{cases}$$

6) In order to have all the poles at -1, we need to solve:

$$s^2 + \left(\frac{L\alpha + JR}{JL} + k_2\frac{\kappa}{JL}\right) s + \left(\frac{R\alpha + \kappa^2}{JL} + k_1\frac{\kappa}{JL}\right) = (s+1)^2$$

or alternatively:

$$\begin{cases} \frac{R\alpha+\kappa^2}{JL} + k_1\frac{\kappa}{JL} = 1 \\ \frac{L\alpha+JR}{JL} + k_2\frac{\kappa}{JL} = 2 \end{cases}$$

and thus:

$$\begin{cases} k_1 = \frac{JL - R\alpha - \kappa^2}{\kappa} \\ k_2 = \frac{2JL - L\alpha - JR}{\kappa} \end{cases}$$

7) At equilibrium, $\dot{\mathbf{x}} = \mathbf{0}$ and therefore:

$$\begin{cases} \mathbf{0} = \begin{pmatrix} 0 & 1 \\ -1 & -2 \end{pmatrix} \bar{\mathbf{x}} + \begin{pmatrix} 0 \\ \frac{\kappa h}{JL} \end{pmatrix} \bar{w} \\ \bar{y} = \begin{pmatrix} 1 & 0 \end{pmatrix} \bar{\mathbf{x}} \end{cases}$$

We thus have:

$$\bar{y} = \begin{pmatrix} 1 & 0 \end{pmatrix} \begin{pmatrix} 0 & 1 \\ -1 & -2 \end{pmatrix}^{-1} \begin{pmatrix} 0 \\ -\frac{\kappa h}{JL} \end{pmatrix} \bar{w}$$

We will have $\bar{w} = \bar{y}$ if:

$$\underbrace{\begin{pmatrix} 1 & 0 \end{pmatrix} \begin{pmatrix} 0 & 1 \\ -1 & -2 \end{pmatrix}^{-1} \begin{pmatrix} 0 \\ -\frac{\kappa h}{JL} \end{pmatrix}}_{= \frac{\kappa}{LJ} h} = 1$$

in other words if:

$$h = \frac{JL}{\kappa}$$

Finally, the control is expressed by:

$$u = \frac{JL}{\kappa} w - \frac{JL - R\alpha - \kappa^2}{\kappa} y - \frac{2JL - L\alpha - JR}{\kappa} \dot{y}$$

Solution to Exercise 4.13 (proportional, integral and derivative control)

1) Since $\mathbf{x} = (y \ \dot{y})^{\mathrm{T}}$, the state equations are written as:

$$\begin{cases} \dot{x}_1 = x_2 \\ \dot{x}_2 = -a_1 x_2 - a_0 x_1 + u \\ y = x_1 \end{cases}$$

2) We have:

$$u = \alpha_{-1}z + \alpha_0(w - x_1) + \alpha_1(\dot{w} - \dot{x}_1) = \alpha_{-1}z + \alpha_0 w$$
$$-\alpha_0 x_1 - \alpha_1 x_2$$

Since $z(t) = \int_0^t e(\tau)\,d\tau$, we have $\dot{z} = -x_1 + w$. The state equations of the controller are therefore:

$$\dot{z} = -x_1 + w$$
$$u = \alpha_{-1}z + \alpha_0 w - \alpha_0 x_1 - \alpha_1 x_2.$$

3) The wiring diagram of the looped system is given in Figure 4.11.

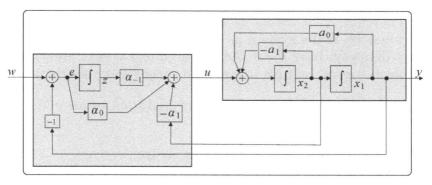

Figure 4.11. *PID controller*

4) We have:

$$\dot{x}_2 = -a_1 x_2 - a_0 x_1 + \alpha_{-1} z + \alpha_0 w - \alpha_0 x_1 - \alpha_1 x_2$$
$$= (-a_0 - \alpha_0) x_1 - (\alpha_1 + a_1) x_2 + \alpha_{-1} z + \alpha_0 w$$

The state equations of the closed-loop system are therefore:

$$\begin{pmatrix} \dot{x}_1 \\ \dot{x}_2 \\ \dot{z} \end{pmatrix} = \begin{pmatrix} 0 & 1 & 0 \\ -a_0 - \alpha_0 & -\alpha_1 - a_1 & \alpha_{-1} \\ -1 & 0 & 0 \end{pmatrix} \begin{pmatrix} x_1 \\ x_2 \\ z \end{pmatrix} + \begin{pmatrix} 0 \\ \alpha_0 \\ 1 \end{pmatrix} w$$

$$y = x_1$$

5) The characteristic polynomial is $P(s) = s^3 + (a_1 + \alpha_1)s^2 + (a_0 + \alpha_0)s + \alpha_{-1}$. In order to have poles at -1, we need to have $P(s) = (s+1)^3 = s^3 + 3s^2 + 3s + 1$. Therefore:

$$\alpha_1 = 3 - a_1, \; \alpha_0 = 3 - a_0, \; \alpha_{-1} = 1$$

6) At equilibrium, we have:

$$\begin{cases} \begin{pmatrix} 0 & 1 & 0 \\ -a_0' - \alpha_0 & -\alpha_1 - a_1' & \alpha_{-1} \\ -1 & 0 & 0 \end{pmatrix} \begin{pmatrix} \bar{x}_1 \\ \bar{x}_2 \\ \bar{z} \end{pmatrix} + \begin{pmatrix} 0 \\ \alpha_0 \bar{w} \\ \bar{w} \end{pmatrix} = \begin{pmatrix} 0 \\ 0 \\ 0 \end{pmatrix} \\ \bar{y} - \bar{x}_1 = 0 \end{cases}$$

Therefore $\bar{y} = \bar{x}_1 = \bar{w}$. Regardless of the value of the parameters, if the system is stable, given the integral term, we will always have $\bar{y} = \bar{w}$. Adding an integral term allows for a robust control relative to any kind of constant disturbance.

Solution to Exercise 4.14 (output feedback of an order 3 system)

1) The state equations of the system are:

$$\begin{cases} \begin{pmatrix} \dot{x}_1 \\ \dot{x}_2 \\ \dot{x}_3 \end{pmatrix} = \begin{pmatrix} 0 & 1 & 0 \\ 0 & 0 & 1 \\ 1 & -4 & 7 \end{pmatrix} \begin{pmatrix} x_1 \\ x_2 \\ x_3 \end{pmatrix} + \begin{pmatrix} 0 \\ 0 \\ 1 \end{pmatrix} u \\ y = \begin{pmatrix} -2 & 7 & 3 \end{pmatrix} \begin{pmatrix} x_1 \\ x_2 \\ x_3 \end{pmatrix} \end{cases}$$

The evolution matrix is a companion matrix. We can directly deduce from the last row of the evolution matrix $\mathbf{A}$ that the characteristic polynomial is $P(s) = s^3 - 7s^2 + 4s - 1$.

2) The looped system verifies:

$$\dot{\mathbf{x}} = \mathbf{A}\mathbf{x} + \mathbf{B}\left(-\mathbf{K}\mathbf{x} + hw\right) = \left(\mathbf{A} - \mathbf{B}\mathbf{K}\right)\mathbf{x} + \mathbf{B}hw$$

with: $\mathbf{K} = (k_1,\ k_2,\ k_3)$. Thus, the evolution matrix of the looped system is:

$$\mathbf{A} - \mathbf{B}\mathbf{K} = \begin{pmatrix} 0 & 1 & 0 \\ 0 & 0 & 1 \\ 1 & -4 & 7 \end{pmatrix} - \begin{pmatrix} 0 \\ 0 \\ 1 \end{pmatrix} \begin{pmatrix} k_1 & k_2 & k_3 \end{pmatrix}$$

$$= \begin{pmatrix} 0 & 1 & 0 \\ 0 & 0 & 1 \\ -k_1+1 & -k_2-4 & -k_3+7 \end{pmatrix}$$

Its characteristic polynomial is:

$$P(s) = s^3 + (k_3 - 7)\,s^2 + (k_2 + 4)\,s + (k_1 - 1)$$

We would like it to be equal to:

$$P_0(s) = (s+1)^3 = s^3 + 3s^2 + 3s + 1$$

By identification, we obtain:

$$\mathbf{K} = \begin{pmatrix} k_1 & k_2 & k_3 \end{pmatrix} = \begin{pmatrix} 2 & -1 & 10 \end{pmatrix}$$

3) At equilibrium, we have:

$$\begin{cases} (\mathbf{A} - \mathbf{B}\mathbf{K})\,\bar{\mathbf{x}} + \mathbf{B}h\bar{w} = 0 \\ \bar{y} = \mathbf{C}\bar{\mathbf{x}} \end{cases}$$

Since $\bar{\mathbf{x}} = -\left(\mathbf{A} - \mathbf{B}\mathbf{K}\right)^{-1}\mathbf{B}h\bar{w}$, we obtain:

$$\bar{y} = -\mathbf{C}\left(\mathbf{A} - \mathbf{B}\mathbf{K}\right)^{-1}\mathbf{B}h\bar{w}$$

In order to have $\bar{y} = \bar{w}$, we need to choose:

$$h = -\left(\mathbf{C}\left(\mathbf{A} - \mathbf{BK}\right)^{-1}\mathbf{B}\right)^{-1}$$

i.e.:

$$h = -\left(\begin{pmatrix}-2 & 7 & 3\end{pmatrix}\begin{pmatrix}0 & 1 & 0 \\ 0 & 0 & 1 \\ -1 & -3 & -3\end{pmatrix}^{-1}\begin{pmatrix}0 \\ 0 \\ 1\end{pmatrix}\right)^{-1} = -\frac{1}{2}$$

Let us note that, for the calculation of h, we do not need to invert the 3×3 matrix, but calculate the intermediate quantity:

$$\mathbf{v} = \begin{pmatrix}0 & 1 & 0 \\ 0 & 0 & 1 \\ -1 & -3 & -3\end{pmatrix}^{-1}\begin{pmatrix}0 \\ 0 \\ 1\end{pmatrix}$$

which in this case can be done mentally. Indeed, it is sufficient to solve:

$$\begin{pmatrix}0 & 1 & 0 \\ 0 & 0 & 1 \\ -1 & -3 & -3\end{pmatrix}\mathbf{v} = \begin{pmatrix}0 \\ 0 \\ 1\end{pmatrix} \Rightarrow \mathbf{v} = \begin{pmatrix}-1 \\ 0 \\ 0\end{pmatrix}$$

Solution to Exercise 4.15 (state feedback with integral effect and monovariate case)

1) The characteristic polynomial of the system is:

$$P(s) = \det\left(s\begin{pmatrix}1 & 0 \\ 0 & 1\end{pmatrix} - \begin{pmatrix}1 & 1 \\ 0 & 2\end{pmatrix}\right)$$

$$= \det\begin{pmatrix}s-1 & -1 \\ 0 & s-2\end{pmatrix} = (s-1)(s-2)$$

The system is unstable, since the poles are positive.

2) The state equations of the controller are:

$$\begin{cases} \dot{z} = w - x_1 \\ u = -k_1 x_1 - k_2 x_2 + \alpha z \end{cases}$$

The only pole of the controller is 0 (but in practice, this has no importance).

3) The looped system has as state equations:

$$\begin{cases} \dot{x}_1 = x_1 + x_2 \\ \dot{x}_2 = 2x_2 - k_1 x_1 - k_2 x_2 + \alpha z \\ \dot{z} = w - x_1 \\ y = x_1 \end{cases}$$

or, in matrix form:

$$\begin{pmatrix} \dot{x}_1 \\ \dot{x}_2 \\ \dot{z} \end{pmatrix} = \begin{pmatrix} 1 & 1 & 0 \\ -k_1 & 2-k_2 & \alpha \\ -1 & 0 & 0 \end{pmatrix} \begin{pmatrix} x_1 \\ x_2 \\ z \end{pmatrix} + \begin{pmatrix} 0 \\ 0 \\ 1 \end{pmatrix} w$$

4) The characteristic polynomial is:

$$P(s) = s^3 + (k_2 - 3) s^2 + (k_1 - k_2 + 2) s + \alpha$$

We would like it to be equal to $s^3 + 3s^2 + 3s + 1$. Whence $\alpha = 1, k_1 = 7, k_2 = 6$.

5) At equilibrium, $\dot{x}_1 = \dot{x}_2 = \dot{z} = 0$. Therefore:

$$\begin{pmatrix} 1 & 1 & 0 \\ -7 & -4 & 1 \\ -1 & 0 & 0 \end{pmatrix} \begin{pmatrix} \bar{x}_1 \\ \bar{x}_2 \\ \bar{z} \end{pmatrix} + \begin{pmatrix} 0 \\ 0 \\ 1 \end{pmatrix} \bar{w} = 0$$

Thus:

$$\begin{pmatrix} \bar{x}_1 \\ \bar{x}_2 \\ \bar{z} \end{pmatrix} = \begin{pmatrix} 1 & 1 & 0 \\ -7 & -4 & 1 \\ -1 & 0 & 0 \end{pmatrix}^{-1} \begin{pmatrix} 0 \\ 0 \\ -\bar{w} \end{pmatrix} = \begin{pmatrix} \bar{w} \\ -\bar{w} \\ 3\bar{w} \end{pmatrix}$$

Thus, $\bar{y} = \bar{x}_1 = \bar{w}$. Let us note that if this is not the case, in other words if $\bar{w} - \bar{x}_1$ was non-zero, then the integrator would not be in equilibrium. This is not possible since the system is stable. The role of the integrator is precisely to ensure zero static error and this, even when there are constant disturbances.

6) If we slightly move the matrices of the system, then, since the eigenvalues of the looped system evolve in a continuous manner depending on these matrices, the roots will remain around -1 and therefore the system will remain stable. Thus, $\dot{z}$ (as well as $\dot{x}_1$ and $\dot{x}_2$) will converge toward 0 (otherwise, z would diverge). Since $\dot{z} = w - y$, $y(t)$ will converge toward $\bar{w}$.

Solution to Exercise 4.16 (state feedback with integral effect, general case)

1) The state equations of the looped system are written as:

$$\begin{cases} \dot{\mathbf{x}} = (\mathbf{A} - \mathbf{BK})\mathbf{x} + \mathbf{BK}_i\mathbf{z} + \mathbf{p} \\ \dot{\mathbf{z}} = \mathbf{w} - \mathbf{Ex} \end{cases}$$

The requested dimensions are $\mathbf{w}: m$, $\mathbf{x}_c: m$, $\mathbf{p}: n$, $\mathbf{z}: m$, $\mathbf{A}: n \times n$, $\mathbf{B}: n \times m$, $\mathbf{E}: m \times n$, $\mathbf{K}_i: m \times m$ and $\mathbf{K}: m \times n$.

2) In matrix form, the equations are written as:

$$\begin{pmatrix} \dot{x} \\ \dot{z} \end{pmatrix} = \begin{pmatrix} A - BK & BK_i \\ -E & 0 \end{pmatrix} \begin{pmatrix} x \\ z \end{pmatrix} + \begin{pmatrix} p \\ w \end{pmatrix}$$

$$\left(\begin{pmatrix} A & 0 \\ -E & 0 \end{pmatrix} - \begin{pmatrix} B \\ 0 \end{pmatrix} \begin{pmatrix} K & K_i \end{pmatrix} \right) \begin{pmatrix} x \\ z \end{pmatrix} + \begin{pmatrix} p \\ w \end{pmatrix}$$

The choice of K and K_i will be done by pole placement. We will need to write, in MATLAB:

```
A1=[A,zeros(n,m);-E,zeros(m,m)];B1=[B;
zeros(m,m)]; place(A1,B1,-ones(1:n+m));
```

3) For a constant setpoint w and disturbance p, in a steady state, we have:

$$\begin{cases} 0 = (A - BK)x + BK_i z + p \\ 0 = w - Ex \end{cases}$$

In other words, $w = Ex = x_c$ and $p = -(A - BK)x - BK_i z$. Thus, the setpoint condition is respected ($x_c = w$) and we are capable to find the disturbance.

4) The state equations of the state feedback controller with integral effect are:

$$\begin{cases} \dot{z} = -Ex + w \\ u = K_i z - Kx \end{cases}$$

Solution to Exercise 4.17 (observer)

1) We have:

$$\begin{cases} \dot{x}_1 = -(u + x_2) + (x_1 + (u + x_2)) = x_1 \\ \dot{x}_2 = (x_1 + (u + x_2)) + (u + x_2) = x_1 + 2x_2 + 2u \\ y = x_2 \end{cases}$$

or, in matrix form:

$$\begin{cases} \begin{pmatrix} \dot{x}_1 \\ \dot{x}_2 \end{pmatrix} = \begin{pmatrix} 1 & 0 \\ 1 & 2 \end{pmatrix} \begin{pmatrix} x_1 \\ x_2 \end{pmatrix} + \begin{pmatrix} 0 \\ 2 \end{pmatrix} u \\ y = x_2 \end{cases}$$

The wiring system is given in Figure 4.12.

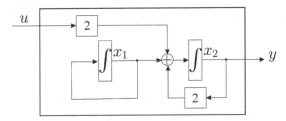

Figure 4.12. *Simplified wiring diagram*

2) The eigenvalues of the system are $\lambda_1 = 1$ and $\lambda_2 = 2$ that have negative real parts. The system is, therefore, unstable.

3) We have:

$$\begin{cases} s\hat{x}_1 = \hat{x}_1 \\ s\hat{x}_2 = \hat{x}_1 + 2\hat{x}_2 + 2\hat{u} \\ \hat{y} = \hat{x}_2 \end{cases} \Rightarrow \begin{cases} \hat{x}_1 = 0 \\ \hat{x}_2 = \frac{2}{s-2}\hat{u} \\ \hat{y} = \frac{2}{s-2}\hat{u} \end{cases}$$

The transfer function is, therefore, $\frac{2}{s-2}$.

4) The controllability matrix is:

$$(\mathbf{B} \mid \mathbf{AB}) = \begin{pmatrix} 0 & 0 \\ 2 & 4 \end{pmatrix}$$

Its rank is < 2, and therefore the system is not controllable. The observability matrix is:

$$\begin{pmatrix} \mathbf{C} \\ \mathbf{CA} \end{pmatrix} = \begin{pmatrix} 0 & 1 \\ 1 & 2 \end{pmatrix}$$

Its rank is equal to 2, and therefore the system is observable.

5) We solve $\det(s\mathbf{I} - \mathbf{A} + \mathbf{LC}) = (s+1)^2$, i.e.:

$$\det\left(\begin{pmatrix} s & 0 \\ 0 & s \end{pmatrix} - \begin{pmatrix} 1 & 0 \\ 1 & 2 \end{pmatrix} + \begin{pmatrix} \ell_1 \\ \ell_2 \end{pmatrix}\begin{pmatrix} 0 & 1 \end{pmatrix}\right) = s^2 + s(-3 + \ell_2)$$
$$+ \ell_1 - \ell_2 + 2 = s^2 + 2s + 1$$

We have $-3 + \ell_2 = 2$ and $\ell_1 - \ell_2 + 2 = 1$. Therefore, $\ell_2 = 5$ and $\ell_1 = 4$. The state equations for the observer are:

$$\begin{cases} \frac{d}{dt}\hat{\mathbf{x}} = (\mathbf{A} - \mathbf{LC})\hat{\mathbf{x}} + \mathbf{B}u + \mathbf{L}y \\ \hat{\mathbf{x}} = \hat{\mathbf{x}} \end{cases}$$

i.e.:

$$\begin{cases} \frac{d}{dt}\hat{\mathbf{x}} = \begin{pmatrix} 1 & -4 \\ 1 & -3 \end{pmatrix}\hat{\mathbf{x}} + \begin{pmatrix} 0 & 4 \\ 2 & 5 \end{pmatrix}\begin{pmatrix} u \\ y \end{pmatrix} \\ \hat{\mathbf{x}} = \hat{\mathbf{x}} \end{cases}$$

Solution to Exercise 4.18 (amplitude demodulator)

1) Since $x_1 = \omega y$, $x_2 = \dot{y}$, we have:

$$\begin{cases} \dot{x}_1 = \omega\dot{y} = \omega x_2 \\ \dot{x}_2 = \ddot{y} = -\omega^2 a\cos(\omega t + b) = -\omega^2 y = -\omega x_1 \end{cases}$$

and therefore the state equations are:

$$\begin{cases} \dot{\mathbf{x}} = \begin{pmatrix} 0 & \omega \\ -\omega & 0 \end{pmatrix} \mathbf{x} \\ y = \begin{pmatrix} \frac{1}{\omega} & 0 \end{pmatrix} \mathbf{x} \end{cases}$$

2) We have:

$$\begin{cases} x_1 = \omega y = a\omega \cos(\omega t + b) \\ x_2 = \dot{y} = -a\omega \sin(\omega t + b) \end{cases}$$

Therefore, $a = \frac{1}{\omega} \|\mathbf{x}\|$.

3) We take a *Luenberger*-type observer given by:

$$\frac{d}{dt}\hat{\mathbf{x}} = (\mathbf{A} - \mathbf{LC})\hat{\mathbf{x}} + \mathbf{L}y$$

We solve:

$$\det(s\mathbf{I} - \mathbf{A} + \mathbf{LC}) = (s+1)^2 = s^2 + 2s + 1$$

However:

$$\det(s\mathbf{I} - \mathbf{A} + \mathbf{LC}) = \det\left(\begin{pmatrix} s & -\omega \\ \omega & s \end{pmatrix} + \begin{pmatrix} \ell_1 \\ \ell_2 \end{pmatrix} \begin{pmatrix} \frac{1}{\omega} & 0 \end{pmatrix}\right)$$

$$= \det\begin{pmatrix} s + \frac{1}{\omega}\ell_1 & -\omega \\ \omega + \frac{1}{\omega}\ell_2 & s \end{pmatrix}$$

Therefore, the equation to solve becomes:

$$s^2 + \frac{1}{\omega}\ell_1 s + \omega^2 + \ell_2 = s^2 + 2s + 1$$

Let $\ell_1 = 2\omega$ and $\ell_2 = 1 - \omega^2$. The observer is therefore:

$$\frac{d}{dt}\hat{\mathbf{x}} = \left(\begin{pmatrix} 0 & \omega \\ -\omega & 0 \end{pmatrix} - \begin{pmatrix} 2\omega \\ 1-\omega^2 \end{pmatrix}\begin{pmatrix} \frac{1}{\omega} & 0 \end{pmatrix}\right)\hat{\mathbf{x}} + \begin{pmatrix} 2\omega \\ 1-\omega^2 \end{pmatrix} y$$

i.e.:

$$\frac{d}{dt}\hat{\mathbf{x}} = \begin{pmatrix} -2 & \omega \\ -\frac{1}{\omega} & 0 \end{pmatrix}\hat{\mathbf{x}} + \begin{pmatrix} 2\omega \\ 1-\omega^2 \end{pmatrix} y$$

4) This filter is described by the following state equations:

$$\begin{cases} \frac{d}{dt}\hat{\mathbf{x}} = \begin{pmatrix} -2 & \omega \\ -\frac{1}{\omega} & 0 \end{pmatrix}\hat{\mathbf{x}} + \begin{pmatrix} 2\omega \\ 1-\omega^2 \end{pmatrix} y \\ \hat{a} = \frac{1}{\omega}\|\hat{\mathbf{x}}\| \end{cases}$$

This type of filter is used to detect the presence of a signal of known frequency (for example, sound of an underwater transponder). If a varies slightly in time, the estimator can also be used to perform amplitude demodulation.

5) The program, which can be found in demodul.m, is given below:

```
tmax=10; dt=0.01; w=1; t=0:dt:tmax;
y=2*cos(w*t)+0.1*randn(size(t));
xhat=[0;0]; yhat=0*t; ahat=0*t;
for k=1:length(t),
xhat=xhat+dt*([-2,w;-1/w 0]*xhat+[2*w;1-w^2]*y(k));
ahat(k)=(1/w)*norm(xhat);
yhat(k)=xhat(1)/w;
end
```

Figure 4.13 illustrates the amplitude demodulation performed by this program. Let us note that $\hat{a}$ depends on time, since as time goes by more information is available to the filter allowing it to refine its estimation.

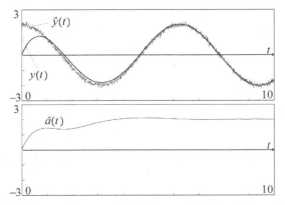

Figure 4.13. *Estimation of the amplitude a of a noisy sinusoidal signal of known frequency*

6) In order to draw the Bode plot in MATLAB, we build the system using the `ss(A,B,C,D)` instruction, then draw the plot using the `bode` instruction.

```
sys = ss([-2,w;-1/w 0],[2*w;1-w^2],[1/w,0],0);
bode(sys);
```

We obtain Figure 4.14.

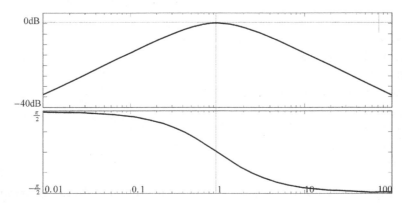

Figure 4.14. *Bode plot of the filter*

We may recognize the frequency response of a bandpass filter centered around the pulsation $\omega = 1$ rad/second.

Solution to Exercise 4.19 (output feedback of a non-strictly proper system)

1) We can proceed by identification. The state equation of $\mathcal{S}$ is of the form:

$$\begin{cases} \dot{x} = ax + bu \\ y = cx + du \end{cases}$$

Its transfer function is:

$$\frac{s+3}{s-2} = \frac{s-2+5}{s-2} = \frac{5}{s-2} + 1 = c(s-a)^{-1}b + d = \frac{cb}{s-a} + d$$

We may take, for instance, $a = 2, b = 5, c = 1, d = 1$. Let us recall that the state representation is not unique.

2) If we take $z = y - du$, we obtain the system described by the equations:

$$\begin{cases} \dot{x} = ax + bu \\ z = cx \end{cases}$$

3) We solve:

$$\begin{cases} s - a + bk = s + 1 \\ s - a + \ell c = s + 1 \end{cases} \Rightarrow \begin{cases} k = \frac{1+a}{b} = \frac{1+2}{5} = \frac{3}{5} \\ \ell = \frac{1+a}{c} = \frac{1+2}{1} = 3 \end{cases}$$

For the precompensator, we have:

$$h = -\left((a-bk)^{-1}b\right)^{-1} = -\left((2-3)^{-1} \cdot 5\right)^{-1} = \frac{1}{5}$$

The equations of the controller are therefore:

$$\begin{cases} \frac{d}{dt}\hat{x} = (a - bk - \ell c)\hat{x} + bhw + \ell z = -4\hat{x} + w + 3z \\ u = -k\hat{x} + hw = -\frac{3}{5}\hat{x} + \frac{1}{5}w \end{cases}$$

4) We have, for the initial system:

$$z = y - u = y + \frac{3}{5}\hat{x} - \frac{1}{5}w$$

Therefore, the controller for $\mathcal{S}$ is:

$$\begin{cases} \frac{d}{dt}\hat{x} = -4\hat{x} + w + 3\left(y + \frac{3}{5}\hat{x} - \frac{1}{5}w\right) = -\frac{11}{5}\hat{x} + \frac{2}{5}w + 3y \\ u = -\frac{3}{5}\hat{x} + \frac{1}{5}w \end{cases}$$

5) The state equations of the looped system are:

$$\begin{cases} \dot{x} = 2x + 5u = 2x - 3\hat{x} + w \\ \frac{d}{dt}\hat{x} = -4\hat{x} + w + 3z = -4\hat{x} + w + 3(x + u - u) \\ \phantom{\frac{d}{dt}\hat{x}} = -4\hat{x} + 3x + w \end{cases}$$

i.e.:

$$\frac{d}{dt}\begin{pmatrix} x \\ \hat{x} \end{pmatrix} = \begin{pmatrix} 2 & -3 \\ 3 & -4 \end{pmatrix} \begin{pmatrix} x \\ \hat{x} \end{pmatrix} + \begin{pmatrix} 1 \\ 1 \end{pmatrix} w$$

We verify that the eigenvalues of the evolution matrix are both equal to -1.

6) Since $y = x + u = x - \frac{3}{5}\hat{x} + \frac{1}{5}w$, the transfer function of the looped system is:

$$G(s) = \mathbf{C}(s\mathbf{I} - \mathbf{A})^{-1}\mathbf{B} + d = \begin{pmatrix} 1 & -\frac{3}{5} \end{pmatrix} \begin{pmatrix} s-2 & 3 \\ -3 & s+4 \end{pmatrix}^{-1} \begin{pmatrix} 1 \\ 1 \end{pmatrix} + \frac{1}{5}$$

$$= \frac{s+3}{5(s+1)}$$

Let us note that a pole-zero simplification has occurred. This simplification comes from the fact that the state $\varepsilon_x = \hat{x} - x$ is a non-controllable state variable and that every non-controllable or non-observable state variable leads to a pole-zero simplification during the calculation of the transfer function.

7) The static gain of the looped system is equal to $\frac{3}{5}$. It is obtained by taking $G(s)$ for $s = 0$. It can also be obtained from the differential equation of the looped system:

$$5\dot{y} + 5y = \dot{w} + 3w$$

At equilibrium, we have $5\bar{y} = 3\bar{w}$. In order to have a static gain equal to 1, we need to add a precompensator before the system with gain $h = \frac{5}{3}$.

Solution to Exercise 4.20 (output feedback with integral effect)

NOTE.– A logical approach would be to express the system in the form:

$$\frac{d}{dt}\begin{pmatrix} \mathbf{x} \\ \mathbf{p} \end{pmatrix} = \begin{pmatrix} \mathbf{A} & \mathbf{I} \\ \mathbf{0} & \mathbf{0} \end{pmatrix} \begin{pmatrix} \mathbf{x} \\ \mathbf{p} \end{pmatrix} + \begin{pmatrix} \mathbf{B} \\ \mathbf{0} \end{pmatrix} \mathbf{u}$$

by including the disturbance vector as a component of the particular state vector. However, the system is not controllable (we cannot control the disturbance) and the pole placement method will not work.

1) The state $(\mathbf{x}, \mathbf{z}, \hat{\mathbf{x}})$ of the looped system is composed of the state x of the system to control, of the vector z of the

memorized quantities in the integrators and of the state $\hat{x}$ reconstructed by the observer. Since $u = -K_i z - K\hat{x}$, we have:

$$\begin{cases} \dot{x} = Ax + B(-K_i z - K\hat{x}) + p \\ \dot{z} = w - ECx \\ \frac{d}{dt}\hat{x} = A\hat{x} + B(-K_i z - K\hat{x}) - L(C\hat{x} - Cx) \end{cases}$$

or, in matrix form:

$$\frac{d}{dt}\begin{pmatrix} x \\ z \\ \hat{x} \end{pmatrix} = \begin{pmatrix} A & -BK_i & -BK \\ -EC & 0 & 0 \\ LC & -BK_i & A - BK - LC \end{pmatrix} \begin{pmatrix} x \\ z \\ \hat{x} \end{pmatrix} + \begin{pmatrix} p \\ w \\ 0 \end{pmatrix}$$

2) We have:

$$\dot{\varepsilon} = \frac{d}{dt}(\hat{x} - x) = LCx - BK_i z + (A - BK - LC)\hat{x} - Ax + BK_i z$$
$$+ BK\hat{x} - p$$
$$= A(\hat{x} - x) - LC(\hat{x} - x) - p = (A - LC)\varepsilon - p$$

Thus:

$$\frac{d}{dt}\begin{pmatrix} x \\ z \\ \varepsilon \end{pmatrix} = \begin{pmatrix} A - BK & -BK_i & -BK \\ -EC & 0 & 0 \\ 0 & 0 & A - LC \end{pmatrix} \begin{pmatrix} x \\ z \\ \varepsilon \end{pmatrix} + \begin{pmatrix} p \\ w \\ -p \end{pmatrix}$$

3) The evolution matrix of the looped system is block diagonal. Thus, the poles of the looped system are composed of the eigenvalues of the following matrix:

$$\begin{pmatrix} A - BK & -BK_i \\ -EC & 0 \end{pmatrix} = \underbrace{\begin{pmatrix} A & 0 \\ -EC & 0 \end{pmatrix}}_{\bar{A}} - \underbrace{\begin{pmatrix} B \\ 0 \end{pmatrix}}_{\bar{B}} \underbrace{\begin{pmatrix} K & K_i \end{pmatrix}}_{\bar{K}}$$

and of those of the matrix $\mathbf{A} - \mathbf{LC}$. We can thus set the poles by doing:

$$\bar{\mathbf{K}} = \texttt{place}(\bar{\mathbf{A}}, \bar{\mathbf{B}}, \mathbf{p}_{\text{con}}), \quad \mathbf{L} = (\texttt{place}(\mathbf{A}^{\mathrm{T}}, \mathbf{C}^{\mathrm{T}}, \mathbf{p}_{\text{obs}}))^{\mathrm{T}}$$

where $\mathbf{p}_{\text{con}}$ and $\mathbf{p}_{\text{obs}}$ are the desired poles.

4) At equilibrium, we have $\dot{\mathbf{z}} = -\mathbf{ECx} + \mathbf{w} = \mathbf{0}$. Therefore:

$$\mathbf{w} = \mathbf{ECx} = \mathbf{y}_c$$

Thus, at equilibrium, the output $\mathbf{y}_c$ is necessarily equal to the setpoint $\mathbf{w}$.

5) The state equations of the controller are:

$$\begin{cases} \dot{\mathbf{z}} = \mathbf{w} - \mathbf{Ey} \\ \frac{d}{dt}\hat{\mathbf{x}} = \mathbf{A}\hat{\mathbf{x}} + \mathbf{B}\left(-\mathbf{K}_i\mathbf{z} - \mathbf{K}\hat{\mathbf{x}}\right) - \mathbf{L}(\mathbf{C}\hat{\mathbf{x}} - \mathbf{y}) \\ \mathbf{u} = -\mathbf{K}_i\mathbf{z} - \mathbf{K}\hat{\mathbf{x}} \end{cases}$$

or alternatively:

$$\begin{cases} \frac{d}{dt}\begin{pmatrix} \mathbf{z} \\ \hat{\mathbf{x}} \end{pmatrix} = \begin{pmatrix} \mathbf{0} & \mathbf{0} \\ -\mathbf{B}\mathbf{K}_i\mathbf{z} & \mathbf{A} - \mathbf{B}\mathbf{K} - \mathbf{LC} \end{pmatrix} \begin{pmatrix} \mathbf{z} \\ \hat{\mathbf{x}} \end{pmatrix} + \begin{pmatrix} \mathbf{I} & -\mathbf{E} \\ \mathbf{0} & \mathbf{L} \end{pmatrix} \begin{pmatrix} \mathbf{w} \\ \mathbf{y} \end{pmatrix} \\ \mathbf{u} \quad = \begin{pmatrix} -\mathbf{K}_i & -\mathbf{K} \end{pmatrix} \begin{pmatrix} \mathbf{z} \\ \hat{\mathbf{x}} \end{pmatrix} \end{cases}$$

5

Linearized Control

In Chapter 4, we have shown how to design controllers for linear systems. However, in practice, the systems are rarely linear. Nevertheless, if their state vector remains localized in a small zone of the state space, the system may be considered linear and the techniques developed in Chapter 4 can then be used. We will first show how to linearize a nonlinear system around a given point of the state space. We will then discuss how to stabilize these nonlinear systems.

5.1. Linearization

5.1.1. *Linearization of a function*

Let $f: \mathbb{R}^n \to \mathbb{R}^p$ be a differentiable function. In the neighborhood of a point $\bar{x} \in \mathbb{R}^n$, the first-order Taylor development of f around $\bar{x}$ gives us:

$$\mathbf{f}(\mathbf{x}) \simeq \mathbf{f}(\bar{\mathbf{x}}) + \frac{d\mathbf{f}}{d\mathbf{x}}(\bar{\mathbf{x}})(\mathbf{x} - \bar{\mathbf{x}})$$

with:

$$\frac{d\mathbf{f}}{d\mathbf{x}}(\bar{\mathbf{x}}) = \begin{pmatrix} \frac{\partial f_1}{\partial x_1}(\bar{\mathbf{x}}) & \frac{\partial f_1}{\partial x_2}(\bar{\mathbf{x}}) & \cdots & \frac{\partial f_1}{\partial x_n}(\bar{\mathbf{x}}) \\ \frac{\partial f_2}{\partial x_1}(\bar{\mathbf{x}}) & \frac{\partial f_2}{\partial x_2}(\bar{\mathbf{x}}) & \cdots & \frac{\partial f_2}{\partial x_n}(\bar{\mathbf{x}}) \\ \vdots & \vdots & & \vdots \\ \frac{\partial f_p}{\partial x_1}(\bar{\mathbf{x}}) & \frac{\partial f_p}{\partial x_2}(\bar{\mathbf{x}}) & \cdots & \frac{\partial f_p}{\partial x_n}(\bar{\mathbf{x}}) \end{pmatrix}$$

This matrix is called the *Jacobian matrix*. Very often, in order to linearize a function, we use formal calculus for calculating the Jacobian matrix, then we instantiate this matrix around $\bar{x}$. When we differentiate by hand, we avoid such proceedings. We save a lot of calculations if we perform the two operations (differentiation and instantiation) simultaneously (see Exercise 5.2). Finally, there is a similar method that is very easy to implement: the *finite difference* method (refer to Exercise 5.4). In order to apply it to the calculation of $\frac{d\mathbf{f}}{d\mathbf{x}}$ at point $\bar{x}$, we approximate the jth column of the Jacobian matrix as follows:

$$\frac{\partial \mathbf{f}}{\partial x_j}(\bar{\mathbf{x}}) \simeq \frac{\mathbf{f}(\bar{\mathbf{x}}+h\mathbf{e}_j) - \mathbf{f}(\bar{\mathbf{x}})}{h}$$

where $\mathbf{e}_j$ is the jth vector of the canonical basis of $\mathbb{R}^n$ and h a small real number. Thus, the Jacobian matrix is approximated by:

$$\frac{d\mathbf{f}}{d\mathbf{x}}(\bar{\mathbf{x}}) \simeq \left(\begin{array}{c|c|c|c} \frac{\mathbf{f}(\bar{\mathbf{x}}+h\mathbf{e}_1)-\mathbf{f}(\bar{\mathbf{x}})}{h} & \frac{\mathbf{f}(\bar{\mathbf{x}}+h\mathbf{e}_2)-\mathbf{f}(\bar{\mathbf{x}})}{h} & \cdots & \frac{\mathbf{f}(\bar{\mathbf{x}}+h\mathbf{e}_n)-\mathbf{f}(\bar{\mathbf{x}})}{h} \end{array} \right)$$

In order for the approximation to be correct, we need to take a very small h. However, if h is too small, numerical problems appear. This method therefore has to be considered weak.

5.1.2. Linearization of a dynamic system

Let us consider the system described by its state equations:

$$\mathcal{S}: \begin{cases} \dot{x} = f(x, u) \\ y = g(x, u) \end{cases}$$

where x is of dimension n, u is of dimension m and y is of dimension p. Around the point $(\bar{x}, \bar{u})$, the behavior of $\mathcal{S}$ is therefore approximated by the following state equations:

$$\begin{cases} \dot{x} = f(\bar{x}, \bar{u}) + A(x - \bar{x}) + B(u - \bar{u}) \\ y = g(\bar{x}, \bar{u}) + C(x - \bar{x}) + D(u - \bar{u}) \end{cases}$$

with:

$$A = \tfrac{\partial f}{\partial x}(\bar{x}, \bar{u}), \ B = \tfrac{\partial f}{\partial u}(\bar{x}, \bar{u})$$

$$C = \tfrac{\partial g}{\partial x}(\bar{x}, \bar{u}), \ D = \tfrac{\partial g}{\partial u}(\bar{x}, \bar{u})$$

This is an affine system called *tangent system* to $\mathcal{S}$ at point $(\bar{x}, \bar{u})$.

5.1.3. Linearization around an operating point

A point $(\bar{x}, \bar{u})$ is an *operating point* (also called *polarization point*) if $f(\bar{x}, \bar{u}) = 0$. If $\bar{u} = 0$, we are in the case of a *point of equilibrium*. Let us note first of all that if $x = \bar{x}$ and if $u = \bar{u}$, then $\dot{x} = 0$, in other words, the system no longer evolves if we maintain the control $u = \bar{u}$ when the system is in the state $\bar{x}$. In this case, the output y has the value $y = \bar{y} = g(\bar{x}, \bar{u})$. Around the operating point, $(\bar{x}, \bar{u})$, the system $\mathcal{S}$ admits the tangent system:

$$\begin{cases} \dot{x} = \quad A(x - \bar{x}) + B(u - \bar{u}) \\ y = \bar{y} + C(x - \bar{x}) + D(u - \bar{u}) \end{cases}$$

Let us take $\tilde{u} = u - \bar{u}$, $\tilde{x} = x - \bar{x}$ and $\tilde{y} = y - \bar{y}$. These vectors are called the *variations* of u, x and y. For small variations $\tilde{u}$, $\tilde{x}$ and $\tilde{y}$, we have:

$$\begin{cases} \frac{d}{dt}\tilde{x} = A\tilde{x} + B\tilde{u} \\ \tilde{y} = C\tilde{x} + D\tilde{u} \end{cases}$$

The system thus formed is called the *linearized system* of $\mathcal{S}$ around the operating point $(\bar{x}, \bar{u})$.

5.2. Stabilization of a nonlinear system

Let us consider the system described by its state equations:

$$\mathcal{S} : \begin{cases} \dot{x} = f(x, u) \\ y = g(x) \end{cases}$$

in which $(\bar{x}, \bar{u})$ constitutes an operating point. In order for our system to behave like a linear system around $(\bar{x}, \bar{u})$, let us create the variables $\tilde{u} = u - \bar{u}$ and $\tilde{y} = y - \bar{y}$, as represented in Figure 5.1.

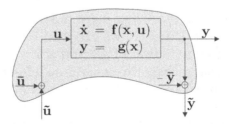

Figure 5.1. *Polarization of the system: the initial system, locally affine (in the neighborhood of the operating point), is transformed into a locally linear system*

The system with input $\widetilde{u}$ and output $\widetilde{y}$ thus designed is called a *polarized system*. The state equations of the polarized system can be approximated by:

$$\begin{cases} \dot{x} = f(x, u) \simeq A.(x - \bar{x}) + B.(u - \bar{u}) = A.(x - \bar{x}) + B.\widetilde{u} \\ \widetilde{y} = -\bar{y} + y = -\bar{y} + g(x) \simeq -\bar{y} + \bar{y} + C(x - \bar{x}) = C(x - \bar{x}) \end{cases}$$

where A, B and C are obtained by the calculation of the Jacobian matrix of the functions f and g at point $(\bar{x}, \bar{u})$. By taking now $\widetilde{x} = x - \bar{x}$ as state vector instead of x, we obtain:

$$\begin{cases} \frac{d}{dt}\widetilde{x} = A\widetilde{x} + B\widetilde{u} \\ \widetilde{y} = C\widetilde{x} \end{cases}$$

which is the linearized system of the nonlinear system. Let $x_c = Ex$ be the sub-vector of the set point variables. We can build a controller $\mathcal{R}_L$ for this linear system using the REGULKLH algorithm (A, B, C, E, p_{con}, p_{obs}) described on page 133. We know that when the set point $\widetilde{w}$ input into $\mathcal{R}_L$ is constant, we have $E\widetilde{x} = \widetilde{w}$. However, we would like the input of the controller w that we are building to satisfy $w = Ex$. We therefore have to build $\widetilde{w}$ from w in such a way that $w = Ex$, at equilibrium. We have:

$$\widetilde{w} = E\widetilde{x} = E(x - \bar{x}) = w - \bar{w}$$

where $\bar{w} = E\bar{x}$. The controller thus obtained is represented in Figure 5.2, in the thick frame.

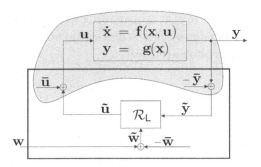

Figure 5.2. *Stabilization of a nonlinear system around an operating point by a linear controller*

This controller stabilizes and decouples our nonlinear system around its operating point. A summary of the method to calculate a controller for a nonlinear system is given below.

Algorithm REGULNL (in: $\mathbf{f}, \mathbf{g}, \mathbf{E}, \mathbf{p}_{\text{con}}, \mathbf{p}_{\text{obs}}, \bar{\mathbf{x}}, \bar{\mathbf{u}}$; out: $\mathcal{R}$)

1 Ensure that $\mathbf{f}(\bar{\mathbf{x}}, \bar{\mathbf{u}}) = \mathbf{0}$;

2 $\bar{\mathbf{y}} := \mathbf{g}(\bar{\mathbf{x}})$; $\bar{\mathbf{w}} := \mathbf{E}\bar{\mathbf{x}}$;

3 $\mathbf{A} := \frac{\partial \mathbf{f}}{\partial \mathbf{x}}(\bar{\mathbf{x}}, \bar{\mathbf{u}})$; $\mathbf{B} := \frac{\partial \mathbf{f}}{\partial \mathbf{u}}(\bar{\mathbf{x}}, \bar{\mathbf{u}})$; $\mathbf{C} := \frac{\partial \mathbf{g}}{\partial \mathbf{x}}(\bar{\mathbf{x}}, \bar{\mathbf{u}})$;

4 $\mathbf{K} := \text{PLACE}(\mathbf{A}, \mathbf{B}, \mathbf{p}_{\text{con}})$;

5 $\mathbf{L} := \left(\text{PLACE}\left(\mathbf{A}^T, \mathbf{C}^T, \mathbf{p}_{\text{obs}}\right)\right)^T$;

6 $\mathbf{H} := -\left(\mathbf{E}\left(\mathbf{A} - \mathbf{B}\mathbf{K}\right)^{-1}\mathbf{B}\right)^{-1}$;

7 $\mathcal{R} := \begin{cases} \frac{d}{dt}\hat{\mathbf{x}} = (\mathbf{A} - \mathbf{B}\mathbf{K} - \mathbf{L}\mathbf{C})\hat{\mathbf{x}} + \mathbf{B}\mathbf{H}(\mathbf{w} - \bar{\mathbf{w}}) + \mathbf{L}(\mathbf{y} - \bar{\mathbf{y}}) \\ \mathbf{u} = \bar{\mathbf{u}} - \mathbf{K}\hat{\mathbf{x}} + \mathbf{H}(\mathbf{w} - \bar{\mathbf{w}}) \end{cases}$

This algorithm returns our controller with inputs y, w and output u in the form of its state equations.

5.3. Exercises

EXERCISE 5.1.– Jacobian matrix

Let us consider the function:

$$f\begin{pmatrix} x_1 \\ x_2 \end{pmatrix} = \begin{pmatrix} x_1^2 x_2 \\ x_1^2 + x_2^2 \end{pmatrix}$$

1) Calculate the Jacobian matrix of **f** at **x**.

2) Linearize this function around $\mathbf{x} = (1, 2)^{\mathrm{T}}$.

EXERCISE 5.2.– Linearization by decomposition

Here, we need to linearize a function $f(\mathbf{x})$ around a point $\bar{\mathbf{x}}$. The decomposition method is used to represent the function f as a composition of functions (sin, exp, etc.) and elementary operators $(+, -, \cdot, /)$. We then calculate, for each intermediary variable u, the values $\partial_i u = \frac{\partial u}{\partial x_i}$ and thus starting from the variables x_i in order to proceed step by step until the function f. The rules that we will be able to apply are of the form:

$$\partial_i (u + v) = \partial_i u + \partial_i v$$

$$\partial_i (u.v) = u.\partial_i v + v.\partial_i u$$

$$\partial_i \left(\frac{u}{v}\right) = \frac{v.\partial_i u - u.\partial_i v}{v^2}$$

$$\partial_i (u^2) = 2u.\partial_i u$$

Linearize the function:

$$f(x_1, x_2) = \frac{\left(x_1^2 + x_1 x_2\right)^2 (x_1 x_2)}{x_1^2 + x_2}$$

around the point $\bar{\mathbf{x}} = (1, 1)^{\mathrm{T}}$ using this approach.

EXERCISE 5.3.– **Linearization of a function by limited development**

One method for the linearization of a function consists of replacing each sub-expression by a limited development (of order 1 or more) around the point of linearization, which we will assume here to be equal to zero. The method allows us to save large amounts of calculation, but it has to be used carefully since it may lead to errors due to singularities. In order to illustrate this principle, we will attempt to linearize the function:

$$f(\mathbf{x}) = \frac{\sin x_1.(3 - 2x_2^2 \cos x_1) + 7 \exp x_2 . \cos x_1}{1 + \sin^2 x_1}$$

around $\bar{\mathbf{x}} = (0,0)$. We have:

$\sin x_1 = x_1 + \varepsilon$

$\cos x_1 = 1 + \varepsilon$

$x_1^2 = x_2^2 = \varepsilon$

$\exp x_2 = 1 + x_2 + \varepsilon$

where $\varepsilon = o\left(\|\mathbf{x}\|\right)$, in other words, that ε is a notation, which means "small before" $\|\mathbf{x}\|$. With this notation, we have rules of the form:

$$(u(\mathbf{x}) + \varepsilon)(v(\mathbf{x}) + \varepsilon) = u(\mathbf{x}).v(\mathbf{x}) + \varepsilon$$

$$\frac{u(\mathbf{x})+\varepsilon}{v(\mathbf{x})+\varepsilon} = \frac{u(\mathbf{x})}{v(\mathbf{x})} + \varepsilon$$

where $u(\mathbf{x})$ and $v(\mathbf{x})$ are two functions with real values, continuous at zero. By using the above stated principle, linearize the function $f(\mathbf{x})$ around zero.

EXERCISE 5.4.– Linearization of a system using the finite difference method in MATLAB

We have the MATLAB code of the evolution function $f(x, u)$ of a system in the form of a script f.m. Give a MATLAB function function[A,B]=Jf(x,u,h) which gives an estimation of the matrices:

$$A = \frac{\partial f}{\partial x}(x, u) \text{ et } B = \frac{\partial f}{\partial u}(x, u)$$

using the finite difference method. Here, h is the step of the method and corresponds to a very small quantity (for example $h = 0.0001$).

EXERCISE 5.5.– Linearization of the predator-prey system

Let us once again consider the Lotka–Volterra predator-prey system, which is given by:

$$\begin{cases} \dot{x}_1 = (1 - x_2) x_1 \\ \dot{x}_2 = (-1 + x_1) x_2 \end{cases}$$

1) Linearize this system around a non-zero point of equilibrium $\bar{x}$.

2) Among the two vector fields of Figure 5.3, one represents the field associated with the Lotka–Volterra system and other corresponds to its tangent system.

3) Calculate the poles of the linearized system. Discuss.

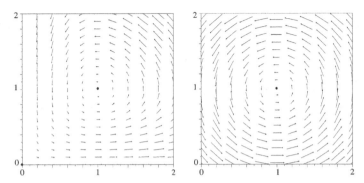

Figure 5.3. *One of the two vector fields is associated with the Lotka–Volterra system. The other corresponds to its tangent system*

EXERCISE 5.6.– **Linearization of a simple pendulum**

The state equations of a simple pendulum are given by:

$$\begin{cases} \begin{pmatrix} \dot{x}_1 \\ \dot{x}_2 \end{pmatrix} = \mathbf{f}(\mathbf{x}, u) = \begin{pmatrix} x_2 \\ \frac{-\ell mg \sin x_1 + u}{m\ell^2} \end{pmatrix} \\ y \quad = \mathbf{g}(\mathbf{x}, u) = \quad \ell \sin x_1 \end{cases}$$

Linearize this pendulum around the point $(\bar{\mathbf{x}} = (0,0), \bar{u} = 0)$.

EXERCISE 5.7.– **Mass in a liquid**

Let us consider a mass in a liquid flow. Its evolution obeys the following differential equation:

$$\ddot{y} + \dot{y}.|\dot{y}| = u$$

where y is the position of the mass and u is the force exerted on this mass.

1) Give the state equation for this system. Linearize around its point of equilibrium. We will take as state vector $\mathbf{x} = (y \ \dot{y})^\mathrm{T}$.

2) Study the stability of the system.

3) In the case of a positive initial velocity $\dot{y}(0)$, calculate the solution of the nonlinear differential equation when $u = 0$. What value does y converge to?

EXERCISE 5.8.– Controllability of the segway

Let us recall that the segway is a vehicle with two wheels and a single axle. We consider first of all that the segway moves in a straight line, and therefore a two-dimensional (2D) model will suffice (see Figure 5.4).

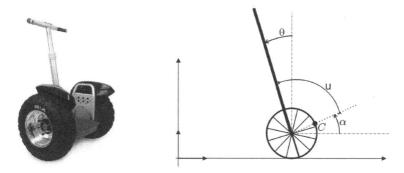

Figure 5.4. *Segway that is the subject of the controllability study*

This system has two degrees of freedom: the angle of the wheel α and the pitch θ. Its state vector is therefore the vector $\mathbf{x} = (\alpha, \theta, \dot{\alpha}, \dot{\theta})^{\mathrm{T}}$. When it is not controlled, the state equations of the segway are (refer to Exercise 1.7):

$$\begin{cases} \begin{pmatrix} \dot{x}_1 \\ \dot{x}_2 \\ \dot{x}_3 \\ \dot{x}_4 \end{pmatrix} = \begin{pmatrix} x_3 \\ x_4 \\ \dfrac{\mu_3\left(\mu_2 x_4^2 - \mu_g \cos x_2\right)\sin x_2 + (\mu_2 + \mu_3 \cos x_2)u}{\mu_1 \mu_2 - \mu_3^2 \cos^2 x_2} \\ \dfrac{\left(\mu_1 \mu_g - \mu_3^2 x_4^2 \cos x_2\right)\sin x_2 - (\mu_1 + \mu_3 \cos x_2)u}{\mu_1 \mu_2 - \mu_3^2 \cos^2 x_2} \end{pmatrix} \\ y \quad = x_1 \end{cases}$$

where:

$$\mu_1 = J_M + a^2(m+M), \quad \mu_2 = J_p + m\ell^2,$$

$$\mu_3 = am\ell, \quad \mu_g = g\ell m$$

The parameters of our system are the mass M of the disk, its radius a, its inertial momentum J_M, the mass m of the body, its inertial momentum J_p and the distance ℓ between its center of gravity and the center of the disk. We have added the observation equation $y = x_1$ in order to assume a situation where only the angle of the wheel α is measured.

1) Calculate the operating points of the system.

2) Linearize the system around an upper point of equilibrium, for $\alpha = 0$.

3) By using the controllability matrix, determine whether the system is controllable.

EXERCISE 5.9.– Control of the segway in MATLAB

Let us consider once again the segway of the previous exercise that has two degrees of freedom: the angle of the wheel α and the pitch θ. Recall that its state vector is therefore the vector $\mathbf{x} = (\alpha, \theta, \dot{\alpha}, \dot{\theta})^{\mathrm{T}}$. We will take $m = 10$ kg, $M = 1$ kg, $\ell = 1$ m, $g = 10$ ms^{-2}, $a = 0.3$ m, $J_p = 10$ kg.m^2 and $J_M = \frac{1}{2}Ma^2$.

1) Simulate this system in a MATLAB environment by using Euler's method, in different situations.

2) We assume that we measure the angle α using an odometer. Calculate an output feedback controller that allows us to place the segway at a given position. We will place all the poles around -2 (be careful, the pole placement algorithm of MATLAB does not allow two identical poles). Test your controller with a non-zero initial condition.

3) Validate the robustness of your control by adding sensor noise to the measure of x, which is white and Gaussian.

4) Given the model of the tank, propose a three-dimensional (3D) model for the segway. The chosen state vector will be:

$$\mathbf{x} = (\theta, \dot{\alpha}, \dot{\theta}, x, y, \psi, \alpha_1, \alpha_2)^{\mathrm{T}}$$

where α_1 is the angle of the right wheel, α_2 is the angle of the left wheel and ψ is the heading of the segway and the (x, y) the coordinates of the center of the segway's axle. The inputs are the body/wheels pair u_1 and the differential between the wheels u_2. Here, $\dot{\alpha}$ represents the average angular velocity between the two wheels, i.e. $\dot{\alpha} = \frac{1}{2}(\dot{\alpha}_1 + \dot{\alpha}_2)$.

5) Propose a control that allows us to regulate the segway, velocity- and heading-wise.

EXERCISE 5.10.– Linearization of the tank

Let us consider the tank system described by the following state equations:

$$\begin{cases} \dot{x} = v \cos \theta \\ \dot{y} = v \sin \theta \\ \dot{\theta} = \omega \\ \dot{\omega} = u_1 \\ \dot{v} = u_2 \end{cases}$$

1) Linearize this system around a state $\bar{\mathbf{x}} = (\bar{x}, \bar{y}, \bar{\theta}, \bar{\omega}, \bar{v})^{\mathrm{T}}$, which is not necessarily an operating point.

2) Study the rank of the controllability matrix at $\bar{\mathbf{x}}$ following the values of $\bar{v}$.

3) Knowing that the columns of the controllability matrix represent the directions in which we can influence the movement and calculate the non-controllable directions.

EXERCISE 5.11.– Linearization of the hovercraft

Let us consider the hovercraft described by the following state equations:

$$\begin{cases} \dot{x} = v_x \\ \dot{y} = v_y \\ \dot{\theta} = \omega \\ \dot{v}_x = u_1 \cos\theta \\ \dot{v}_y = u_1 \sin\theta \\ \dot{\omega} = u_2 \end{cases}$$

1) Linearize around a point $\bar{\mathbf{x}} = (\bar{x}, \bar{y}, \bar{\theta}, \bar{v}_x, \bar{v}_y, \bar{\omega})^{\mathrm{T}}$, which is not necessarily an operating point.

2) Calculate the controllability matrix around $\bar{\mathbf{x}}$.

3) Give the conditions for which this matrix is not of full rank.

4) Knowing that the columns of the controllability matrix represent the directions in which we can influence the movement and calculate the non-controllable directions. Discuss.

EXERCISE 5.12.– Linearization of the single-acting cylinder

Let us consider once more the pneumatic cylinder modeled in Exercise 1.17. Its state equations are given by:

$$\mathcal{S}: \begin{cases} \dot{x}_1 = x_2 \\ \dot{x}_2 = \dfrac{ax_3 - kx_1}{m} \\ \dot{x}_3 = -\dfrac{x_3}{x_1}\left(x_2 - \dfrac{u}{a}\right) \end{cases}$$

1) Let us assume that $x_1 > 0$. Calculate all the possible operating points $(\bar{\mathbf{x}}, \bar{u})$ for $\mathcal{S}$.

2) Now differentiate $\mathcal{S}$ around an operating point $(\bar{\mathbf{x}}, \bar{u})$. Discuss.

EXERCISE 5.13.– Pole placement control of a nonlinear system

Let us consider the nonlinear system given by the state equations:

$$\mathcal{S} : \begin{cases} \dot{x} = 2x^2 + u \\ y = 3x \end{cases}$$

which we wish to stabilize around the state $\bar{x} = 2$. At equilibrium, we would like y to be equal to its set point w. Moreover, we would like all the poles of the looped system to be equal to -1.

1) Give the state equations of the controller using pole placement that satisfies these constraints.

2) What are the state equations of the looped system?

EXERCISE 5.14.– State feedback of a wiring system

Let us consider the system represented by the wiring diagram of Figure 5.5.

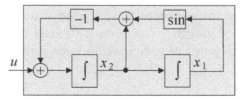

Figure 5.5. *Wiring diagram for a nonlinear system*

1) Give the state equations of the system.

2) Calculate its points of equilibrium.

3) Linearize this system around a point of equilibrium $\bar{x}$ corresponding to $x_1 = \pi$. Is this a stable point of equilibrium?

4) Propose a state feedback controller of the form $u = -\mathbf{K}(\mathbf{x} - \bar{\mathbf{x}})$ that stabilizes the system around $\bar{x}$. We will place the poles at -1.

EXERCISE 5.15.– Controlling an inverted rod pendulum in MATLAB

Let us consider the inverted rod pendulum represented in Figure 5.6, which is composed of a pendulum placed in unstable equilibrium on a carriage.

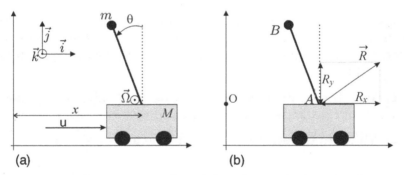

Figure 5.6. *Inverted rod pendulum that we need to control*

The value u is the force exerted on the carriage of mass $M = 5\ kg$, x indicates the position of the carriage, θ is the angle between the pendulum and the vertical axis and $\vec{R}$ is the force exerted by the carriage on the pendulum. At the tip B of the pendulum of length $\ell = 4\ m$ is a fixated mass $m = 1\ Kg$. Finally, A is the point of articulation between the rod and the carriage. As seen in Exercise 1.5, the state equations of this system are:

$$\begin{cases} \begin{pmatrix} \dot{x}_1 \\ \dot{x}_2 \\ \dot{x}_3 \\ \dot{x}_4 \end{pmatrix} = \begin{pmatrix} x_3 \\ x_4 \\ \frac{m \sin x_2 (g \cos x_2 - \ell x_4^2) + u}{M + m \sin^2 x_2} \\ \frac{\sin x_2 ((M+m)g - m\ell x_4^2 \cos x_2) + \cos x_2 u}{\ell (M + m \sin^2 x_2)} \end{pmatrix} \\ y = x_1 \end{cases}$$

where:

$$\mathbf{x} = (x_1, x_2, x_3, x_4)^{\mathrm{T}} = (x, \theta, \dot{x}, \dot{\theta})^{\mathrm{T}}$$

The observation equation indicates that only the position of the carriage x_1 is measured.

1) Calculate all the operating points of the system.

2) Linearize the system around an operating point $\bar{\mathbf{x}} = (0, 0, 0, 0)$ and $\bar{u} = 0$.

3) With MATLAB, obtain an output feedback controller that sets all the poles of the looped system to -2. We propose a control in position, which means that the set point w corresponds to x_1.

4) Create a state feedback control that allows us to raise the pendulum when it leaves its bottom position. For this, we will attempt to stabilize the mechanical energy of the pendulum before moving on to a linear control. We will assume that $u \in [-u_{\max}, u_{\max}]$.

EXERCISE 5.16.– Controlling a car along a wall in MATLAB

Here, we will address the case of a car that drives on an unknown road. The car is equipped with (i) a rangefinder that measures the lateral distance d between the rear axle of the car and the edge of the road, (ii) a velocity sensor that measures the velocity v of the front wheels and (iii) an angle sensor that measures the angle δ of the steering wheel (for reasons of simplification, we will assume that δ also corresponds to the angle between the front wheels and the axis of the car). For our simulations, we will assume that our car is driving around a closed polygon as shown in Figure 5.7.

Figure 5.7. *Car turning around a polygon*

1) Let m, a and b be three points of $\mathbb{R}^2$ and $\overrightarrow{u}$ a vector. Show that the half-line $\mathcal{E}(m, \overrightarrow{u})$ with vertex m and direction vector $\overrightarrow{u}$ intersect the segment [ab] if and only if:

$$\begin{cases} \det(\mathbf{a}-\mathbf{m}, \overrightarrow{u}) \cdot \det(\mathbf{b}-\mathbf{m}, \overrightarrow{u}) \leq 0 \\ \det(\mathbf{a}-\mathbf{m}, \mathbf{b}-\mathbf{a}) \cdot \det(\overrightarrow{u}, \mathbf{b}-\mathbf{a}) \geq 0 \end{cases}$$

Moreover, show that in this case, the distance from the point m to the segment [ab] following the vector $\overrightarrow{u}$ is given by:

$$d = \frac{\det(\mathbf{a}-\mathbf{m}, \mathbf{b}-\mathbf{a})}{\det(\overrightarrow{u}, \mathbf{b}-\mathbf{a})}$$

2) We would like to evolve the car (without using a controller) around the polygon using Euler's method. As seen

in Exercise 1.11, we could take as evolution equation:

$$\begin{pmatrix} \dot{x} \\ \dot{y} \\ \dot{\theta} \\ \dot{v} \\ \dot{\delta} \end{pmatrix} = \begin{pmatrix} v \cos \delta \cos \theta \\ v \cos \delta \sin \theta \\ \frac{v \sin \delta}{L} \\ u_1 \\ u_2 \end{pmatrix}$$

with $L = 3$ m. Here, (x, y) represents the coordinates of the center of the car, θ its heading. The polygon representing the circuit will be represented by a matrix **P** containing 2 rows and $p + 1$ columns. The jth column corresponds to the jth vertex of the polygon. Since the polygon is closed, the first and last columns of **P** are identical. Give a MATLAB function called `circuit_g(x,P)` that corresponds to the observation function of the system and whose arguments are the state of the system x and the polygon P. This function will have to return the distance measured by the rangefinder as well as v and δ. Also, give a MATLAB function called `circuit_f(x,u)` that corresponds to the evolution function.

3) We would like the car to move with a constant velocity along the road. Of course, it is out of the question to involve the shape of the road in the controller, since it is unknown. Moreover, the position and orientation variables of the car are not measured [in other words, we have no compass or Global Positioning System (GPS)]. These quantities, however, are often unnecessary in order to reach the given objective, which is following the road. Indeed, are we not ourselves capable of driving a car on a road without having a local map, without knowing where we are and where North is? What we are interested in when driving a car is the relative position of the car with respect to the road. The world as seen by the controller is represented in Figure 5.8. In this world, the car

moves laterally, like in a computer game, and the edge of the road remains fixed.

This world has to be such that the model used by the controller has state variables describing this relative position of the car with respect to the road and that the constancy of the state variables of this model corresponds to a real situation in which the car follows the road with a constant velocity. We need to imagine an ideal world for the controller in which the associated model admits an operating point that corresponds to the desired behavior of our real system. For this, we will assume that our car drives at a distance $\bar{x}$ of the edge. This model has to have the same inputs u and outputs y as the real system, more precisely two inputs (the acceleration of the front wheels $\dot{v}$ and the angular velocity of the steering wheel $\dot{\delta}$) and three outputs (the distance d of the center of the rear axle to the edge of the road, the velocity v of the front wheels and the angle δ of the steering wheel). Given the fact that only the relative position of the car is of interest to us, our model must only contain four state variables: x, θ, v, δ, as represented in the figure. We must be careful though because the meaning of the variable x has changed since the previous model. Find the state equations (in other words, the evolution equation and the observation equation) of the system imagined by the controller.

4) Let us choose as operating point:

$$\bar{\mathbf{x}} = (5, \pi/2, 7, 0) \text{ and } \bar{\mathbf{u}} = (0, 0)$$

which corresponds to a velocity of 7 ms^{-1} and a distance of 5 m between the center of the rear axle and the edge of the road. Linearize the system around its operating point.

5) Let us take as set point variables x and v, which means that we want our set points w_1 and w_2 to correspond to the distance of the edge of the road and the velocity of the car. Design, in MATLAB, a controller by pole placement. All the

poles will be placed at -2.

6) Test the behavior of your controller in the case where the car is turning around the polygon. Note that the system being controlled (in other words, the turning car) by the controller is different than that it thinks that it is controlling (i.e. a car in a straight line).

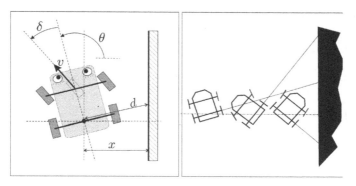

Figure 5.8. *The controller imagines a simpler world than reality, with a straight vertical wall*

EXERCISE 5.17.– Nonlinear control

FIRST PART (proportional and derivative control).– Let us consider a second-order system described by the differential equation $\ddot{y} = u$, where u is the input and y the output.

1) Give a state representation for this system. The chosen state vector is given by $\mathbf{x} = (y, \dot{y})$.

2) We would like to control this system using a proportional and derivative control of the type:

$$u = k_1(w - y) + k_2(\dot{w} - \dot{y}) + \ddot{w}$$

where w is a known function (for instance a polynomial) called *set point* and that can here depend on the time t. Let us note that here, we have assumed that y and $\dot{y}$ were available for the controller. Give the differential equation satisfied by the error $e = w - y$ between the set point and the output y.

3) Calculate k_1 and k_2, allowing to have an error e that converges toward 0 in an exponential manner, with all the poles equal to -1.

SECOND PART (control of a tank-like vehicle around a path).– Let us consider the robotic vehicle on the left-hand side of Figure 5.9, described by the following state equations (tank model):

$$\begin{cases} \dot{x} = v \cos \theta \\ \dot{y} = v \sin \theta \\ \dot{\theta} = u_1 \\ \dot{v} = u_2 \end{cases}$$

where v is the velocity of the robot, θ its orientation and (x, y) the coordinates of its center. We assume that we are capable of measuring all the state variables of our robot with great precision.

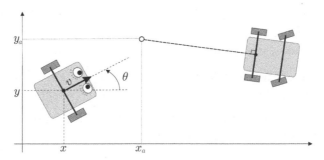

Figure 5.9. *Our robot (with eyes) follows a vehicle (here a car) that has an attachment point (small white circle) to which the robot has to attach*

1) Let us denote by $\mathbf{x} = (x, y, \theta, v)$ the state vector of our robot and $\mathbf{u} = (u_1, u_2)$ the vector of the inputs. Calculate $\ddot{x}$ and $\ddot{y}$ as a function of $\mathbf{x}$ of $\mathbf{u}$. Show that:

$$\begin{pmatrix} \ddot{x} \\ \ddot{y} \end{pmatrix} = \mathbf{A}(\mathbf{x}).\mathbf{u}$$

where $\mathbf{A}(\mathbf{x})$ is a 2×2 matrix whose expression we will give.

2) Another mobile vehicle (on the right-hand side of Figure 5.9) tells us, in a wireless manner, the precise coordinates (x_a, y_a) of a virtual attachment point (in other words, one that only exists mentally), fixed with respect to this vehicle, to which we need to position ourselves. This means that we want the center of the robot (x, y) to be such that $x(t) = x_a(t)$ and $y(t) = y_a(t)$. This vehicle also sends us the first two derivatives $(\dot{x}_a, \dot{y}_a, \ddot{x}_a, \ddot{y}_a)$ of the coordinates of the attachment points. Propose the expression of a control $\mathbf{u}\,(\mathbf{x}, x_a, y_a, \dot{x}_a, \dot{y}_a, \ddot{x}_a, \ddot{y}_a)$, which ensures us that the distance between the position (x, y) of our robot and that of the attachment point (x_a, y_a) decreases in an exponential manner. We will set the poles at -1.

ADVICE.– Perform a first loop as such: $\mathbf{u} = \mathbf{A}^{-1}(\mathbf{x}) \cdot \mathbf{q}$, where $\mathbf{q} = (q_1, q_2)$ is our new input. This first loop will allow us to simplify your system by making it linear. Then, proceed with a second, proportional and derivative loop.

5.4. Solutions

Solution to Exercise 5.1 (Jacobian matrix)

The Jacobian matrix of $\mathbf{f}$ at $\mathbf{x}$ is:

$$\frac{d\mathbf{f}}{d\mathbf{x}}(\mathbf{x}) = \begin{pmatrix} \frac{\partial f_1}{\partial x_1}(\mathbf{x}) & \frac{\partial f_1}{\partial x_2}(\mathbf{x}) \\ \frac{\partial f_2}{\partial x_1}(\mathbf{x}) & \frac{\partial f_2}{\partial x_2}(\mathbf{x}) \end{pmatrix} = \begin{pmatrix} 2x_1 x_2 & x_1^2 \\ 2x_1 & 2x_2 \end{pmatrix}$$

Around the point $\bar{\mathbf{x}} = (1, 2)$, we have:

$$\mathbf{f}\begin{pmatrix} x_1 \\ x_2 \end{pmatrix} \simeq \begin{pmatrix} 2 \\ 5 \end{pmatrix} + \begin{pmatrix} 4 & 1 \\ 2 & 4 \end{pmatrix}\begin{pmatrix} x_1 - 1 \\ x_2 - 2 \end{pmatrix} = \begin{pmatrix} -4 + 4x_1 + x_2 \\ -5 + 2x_1 + 4x_2 \end{pmatrix}$$

Solution to Exercise 5.2 (linearization by decomposition)

We obtain the following table:

$x_1 \doteq 1$	$\partial_1 x_1 = 1$	$\partial_2 x_1 = 0$
$x_2 \doteq 1$	$\partial_1 x_2 = 0$	$\partial_2 x_2 = 1$
$z_1 = x_1^2 \doteq 1$	$\partial_1 z_1 = 2x_1.\partial_1 x_1 \doteq 2$	$\partial_2 z_1 = 0$
$z_2 = x_1 x_2 \doteq 1$	$\partial_1 z_2 = x_2 \doteq 1$	$\partial_2 z_2 = x_1 \doteq 1$
$z_3 = z_1 + z_2 \doteq 2$	$\partial_1 z_3 = \partial_1 z_1 + \partial_1 z_2 \doteq 3$	$\partial_2 z_3 = \partial_2 z_1 + \partial_2 z_2 \doteq 1$
$z_4 = z_3^2 \doteq 4$	$\partial_1 z_4 = 2z_3.\partial_1 z_3 \doteq 12$	$\partial_2 z_4 = 2z_3.\partial_2 z_3 \doteq 4$
$z_5 = z_4 z_2 \doteq 4$	$\partial_1 z_5 = z_4.\partial_1 z_2 + z_2.\partial_1 z_4 \doteq 16$	$\partial_2 z_5 = z_4.\partial_2 z_2 + z_2.\partial_2 z_4 \doteq 8$
$z_6 = z_1 + x_2 \doteq 2$	$\partial_1 z_6 = \partial_1 z_1 + \partial_1 x_2 \doteq 2$	$\partial_2 z_6 = \partial_2 z_1 + \partial_2 x_2 \doteq 1$
$f = \frac{z_5}{z_6} \doteq 2$	$\partial_1 f = \frac{z_6.\partial_1 z_5 - z_5.\partial_1 z_6}{z_6^2} \doteq 6$	$\partial_2 f = \frac{z_6.\partial_2 z_5 - z_5.\partial_2 z_6}{z_6^2} \doteq 3$

Table 5.1. *All intermediate variables with their partial derivatives*

In this table, the symbol $\doteq$ means *is equal after instantiation*. The left-hand column gives the decomposition of the function f into intermediary variables, denoted here by z_i. The second and third columns give the partial derivatives $\frac{\partial}{\partial x_1}$ and $\frac{\partial}{\partial x_2}$. Thus, around $\bar{\mathbf{x}} = (1,1)$, we have the following first-order estimation:

$$f(x_1, x_2) \simeq 2 + 6(x_1 - 1) + 3(x_2 - 1) = 6x_1 + 3x_2 - 7$$

This technique is easily programmed due to the operator overload allowed by many object-oriented programming languages (such as C++). It is at the foundation of automatic differentiation methods allowing us to calculate the differential of a function expressed by computer code.

Solution to Exercise 5.3 (linearization of a function by limited development)

We have:

$$\begin{aligned} f(\mathbf{x}) &= \frac{(x_1+\varepsilon)(3-2\varepsilon(1+\varepsilon))+7(1+x_2+\varepsilon)(1+\varepsilon)}{1+(x_1+\varepsilon)^2} \\ &= \frac{(3x_1+\varepsilon)+7(1+x_2+\varepsilon)}{1+(x_1+\varepsilon)^2} \\ &= \frac{3x_1+7x_2+7+\varepsilon}{1+\varepsilon} \\ &= 3x_1+7x_2+7+\varepsilon \end{aligned}$$

Thus, around $\bar{\mathbf{x}} = (0,0)$, we have the following first-order approximation:

$$f(\mathbf{x}) \simeq 3x_1 + 7x_2 + 7$$

Solution to Exercise 5.4 (linearization of a system using the finite difference method in MATLAB)

In order to have $\mathbf{A} = \frac{\partial \mathbf{f}}{\partial \mathbf{x}}(\bar{\mathbf{x}}, \bar{\mathbf{u}})$ and $\mathbf{B} = \frac{\partial \mathbf{f}}{\partial \mathbf{u}}(\bar{\mathbf{x}}, \bar{\mathbf{u}})$ from $\mathbf{f}, \bar{\mathbf{x}}, \bar{\mathbf{u}}$, without performing any symbolic calculation, we can use a finite difference method in MATLAB. The corresponding MATLAB code is given below.

```
function [A,B]=Jf(x,u,h)
A=[];
for j=1:size(x,1),
dx=0*x; dx(j)=h;
A=[A,1/h*(f(x+dx,u)-f(x,u))];
end;
B=[];
for j=1:size(u,1),
du=0*u; du(j)=h;
B=[B,1/h*(f(x,u+du)-f(x,u))];
end;
```

Solution to Exercise 5.5 (linearization of the predator-prey system)

1) The point of equilibrium is given by:

$$\begin{pmatrix} \bar{x}_1 \\ \bar{x}_2 \end{pmatrix} = \begin{pmatrix} 1 \\ 1 \end{pmatrix}$$

Around this point, $\mathbf{f}(\mathbf{x})$ can be estimated by its tangent system:

$$\mathbf{f}(\mathbf{x}) \simeq \mathbf{f}(\bar{\mathbf{x}}) + \frac{d\mathbf{f}}{d\mathbf{x}}(\bar{\mathbf{x}})(\mathbf{x} - \bar{\mathbf{x}}) = \begin{pmatrix} 1 - \bar{x}_2 & -\bar{x}_1 \\ \bar{x}_2 & -1 + \bar{x}_1 \end{pmatrix} \begin{pmatrix} x_1 - \bar{x}_1 \\ x_2 - \bar{x}_2 \end{pmatrix}$$

$$= \begin{pmatrix} 0 & -1 \\ 1 & 0 \end{pmatrix} \begin{pmatrix} x_1 - 1 \\ x_2 - 1 \end{pmatrix} = \begin{pmatrix} -x_2 + 1 \\ x_1 - 1 \end{pmatrix}$$

The linearized system is obtained by taking $\tilde{x}_1 = x_1 - \bar{x}_1$ and $\tilde{x}_2 = x_2 - \bar{x}_2$, which is given by:

$$\frac{d}{dt}\tilde{\mathbf{x}} = \begin{pmatrix} 0 & -1 \\ 1 & 0 \end{pmatrix} \tilde{\mathbf{x}}$$

2) The field on the left has two points of equilibrium. It corresponds therefore to the Lotka–Volterra system. The one on the right is thus the tangent system at point $(1,1)^{\mathrm{T}}$.

3) The eigenvalues are obtained by calculating the roots of the characteristic polynomial. Since:

$$\det \begin{pmatrix} s & 1 \\ -1 & s \end{pmatrix} = s^2 + 1$$

the eigenvalues are $\pm i$. These values correspond to an oscillating system.

Solution to Exercise 5.6 (linearization of a simple pendulum)

At point ($\bar{x} = (0,0), \bar{u} = 0$), we have $\sin(x_1) = x_1 + \varepsilon$. The linearized system is therefore written as:

$$\begin{cases} \dot{x} = \begin{pmatrix} 0 & 1 \\ -\frac{g}{\ell} & 0 \end{pmatrix} x + \begin{pmatrix} 0 \\ \frac{1}{m\ell^2} \end{pmatrix} u \\ y = \begin{pmatrix} \ell & 0 \end{pmatrix} x \end{cases}$$

Solution to Exercise 5.7 (mass in a liquid)

1) Let us take $x = (y \ \dot{y})^T$. We have the following state equations:

$$\begin{cases} \dot{x}_1 = x_2 \\ \dot{x}_2 = -x_2 \cdot |x_2| + u \end{cases}$$

2) The linearized system is therefore written as:

$$\dot{x} = \begin{pmatrix} 0 & 1 \\ 0 & 0 \end{pmatrix} + \begin{pmatrix} 0 \\ 1 \end{pmatrix} u$$

The two eigenvalues are nil which implies the instability of the linearized system. Let us note, however, that the liquid friction does not appear in the expression of the linearized system. This means that for an input $u = 0$, the velocity x_2 of the linearized system remains unchanged, whereas for the nonlinear system, the velocity x_2 converges toward 0.

3) We will limit ourselves here to a positive initial velocity $\dot{y}(0) = x_2(0)$. The solution of the differential equation is easily calculated. It is given by:

$$y(t) = y(0) + \ln(-\dot{y}(0)) + \ln(t - \dot{y}^{-1}(0))$$

or, equivalently:
$$x_1(t) = x_1(0) + \ln(-x_2(0)) + \ln\left(t - x_2^{-1}(0)\right)$$
$$x_2(t) = \frac{1}{t - x_2^{-1}(0)}$$

Indeed:
$$\ddot{y} = \dot{x}_2 = -\frac{1}{\left(t - x_2^{-1}(0)\right)^2} = -x_2^2 = -\dot{y}^2$$

Not only the linear system but also the linearized one for a non-zero initial velocity x_1 diverges at infinity when $u = 0$.

Solution to Exercise 5.8 (controllability of the segway)

1) The operating points satisfy $\mathbf{f}(\bar{\mathbf{x}}, \bar{\mathbf{u}}) = \mathbf{0}$, in other words:
$$\begin{cases} \bar{x}_3 = 0 \\ \bar{x}_4 = 0 \\ \mu_3\left(\mu_2 \bar{x}_4^2 - \mu_g \cos \bar{x}_2\right) \sin \bar{x}_2 + \left(\mu_2 + \mu_3 \cos \bar{x}_2\right) \bar{u} = 0 \\ \left(\mu_1 \mu_g - \mu_3^2 \bar{x}_4^2 \cos \bar{x}_2\right) \sin \bar{x}_2 - \left(\mu_1 + \mu_3 \cos \bar{x}_2\right) \bar{u} = 0 \end{cases}$$

Since $\bar{x}_4 = 0$, the last two rows are written in the form:
$$\begin{pmatrix} -\mu_g \mu_3 \cos \bar{x}_2 & (\mu_2 + \mu_3 \cos \bar{x}_2) \\ \mu_1 \mu_g & -(\mu_1 + \mu_3 \cos \bar{x}_2) \end{pmatrix} \begin{pmatrix} \sin \bar{x}_2 \\ \bar{u} \end{pmatrix} = \begin{pmatrix} 0 \\ 0 \end{pmatrix}$$

The determinant of the matrix is zero if:
$$\mu_1 \mu_g \mu_3 \cos \bar{x}_2 + \mu_g \mu_3^2 \cos^2 \bar{x}_2 - \mu_1 \mu_g \mu_2 - \mu_1 \mu_g \mu_3 \cos \bar{x}_2 = 0$$

i.e.:
$$\cos^2 \bar{x}_2 = \frac{\mu_1 \mu_2}{\mu_3^2}$$

or:

$$\frac{\mu_1\mu_2}{\mu_3^2} = \frac{(J_M + a^2(m+M))(J_p + m\ell^2)}{(am\ell)^2}$$

$$> \frac{(a^2(m+M))(m\ell^2)}{(am\ell)^2} > 1$$

We can therefore conclude that the determinant of the matrix is never zero and consequently that:

$$\begin{pmatrix} \sin \bar{x}_2 \\ \bar{u} \end{pmatrix} = \begin{pmatrix} 0 \\ 0 \end{pmatrix}$$

The operating points are therefore of the form:

$$\bar{\mathbf{x}} = (\bar{x}_1, k\pi, 0, 0)^T \text{ and } \bar{u} = 0 \text{ with } k \in \mathbb{Z}$$

2) Let us linearize this system around the operating point $\bar{u} = 0$, $\bar{\mathbf{x}} = \mathbf{0}$. We obtain the linearized system described by the following state equations:

$$\dot{\mathbf{x}} = \begin{pmatrix} 0 & 0 & 1 & 0 \\ 0 & 0 & 0 & 1 \\ 0 & -\frac{\mu_3\mu_g}{\mu_1\mu_2-\mu_3^2} & 0 & 0 \\ 0 & \frac{\mu_1\mu_g}{\mu_1\mu_2-\mu_3^2} & 0 & 0 \end{pmatrix} \mathbf{x} + \begin{pmatrix} 0 \\ 0 \\ \frac{\mu_2+\mu_3}{\mu_1\mu_2-\mu_3^2} \\ -\frac{\mu_1+\mu_3}{\mu_1\mu_2-\mu_3^2} \end{pmatrix} u$$

3) The controllability matrix is:

$$\Gamma_{\text{con}} = \begin{pmatrix} 0 & b_3 & 0 & b_4 a_{32} \\ 0 & b_4 & 0 & b_4 a_{42} \\ b_3 & 0 & b_4 a_{32} & 0 \\ b_4 & 0 & b_4 a_{42} & 0 \end{pmatrix}$$

where a_{ij} and b_i correspond to the coefficients of the Jacobian matrices $\mathbf{A}$ and $\mathbf{B}$. The system is uncontrollable if Γ_{con} is not invertible, in other words, if:

$$b_4(b_3 a_{42} - b_4 a_{32}) = 0$$

Since b_4 cannot be nil (since $\mu_1 + \mu_3 > 0$), we have:

$$b_3 a_{42} = b_4 a_{32} \Leftrightarrow \frac{\mu_2+\mu_3}{\mu_1\mu_2-\mu_3^2}\frac{\mu_1\mu_g}{\mu_1\mu_2-\mu_3^2} = \frac{\mu_1+\mu_3}{\mu_1\mu_2-\mu_3^2}\frac{\mu_3\mu_g}{\mu_1\mu_2-\mu_3^2}$$
$$\Leftrightarrow (\mu_2+\mu_3)\mu_1 = (\mu_1+\mu_3)\mu_3$$
$$\Leftrightarrow \mu_2\mu_1 = \mu_3^2$$
$$\Leftrightarrow (J_M + a^2(m+M))(J_p + m\ell^2) = (am\ell)^2$$

which is only possible if $J_M = 0$, $J_p = 0$ and $M = 0$ (which is not realistic). The matrix Γ_{con} is therefore necessarily invertible, which implies that the system is controllable.

Solution to Exercise 5.9 (control of the segway in MATLAB*)*

1) The equations of the segway are:

$$\begin{pmatrix} \dot{x}_1 \\ \dot{x}_2 \\ \dot{x}_3 \\ \dot{x}_4 \end{pmatrix} = \begin{pmatrix} x_3 \\ x_4 \\ \frac{\mu_3(\mu_2 x_4^2 - \mu_g \cos x_2)\sin x_2 + (\mu_2 + \mu_3 \cos x_2)u}{\mu_1\mu_2 - \mu_3^2 \cos^2 x_2} \\ \frac{(\mu_1\mu_g - \mu_3^2 x_4^2 \cos x_2)\sin x_2 - (\mu_1 + \mu_3 \cos x_2)u}{\mu_1\mu_2 - \mu_3^2 \cos^2 x_2} \end{pmatrix}$$

The MATLAB program is given below (without the graphical part).

```
function v=segway2D_f(x,u)
m=10;M=1;l=1;g=10;a=0.3;Jp=10;JM=0.5*M*a^2;
mu1=a^2*(m+M)+JM;  mu2=Jp+m*l^2; mu3=m*a*l;
mug=g*m*l;
c2=cos(x(2));s2=sin(x(2));
den=mu1*mu2-(mu3*c2)^2;
w2=x(4)^2;    v=[x(3);   x(4);
(1/den)*(mu3*(mu2*w2-mug*c2)*s2+(mu2+mu3*c2)*u);
```

```
(1/den)*((mu1*mug-mu3^2*w2*c2)*s2-(mu1+mu3*c2)*u)];
end
```

The main program `segway2D_main.m` is given below.

```
x=[0;0;0;0.01]; % initial state of the segway
dt = 0.01;
m=10;M=1;l=1;g=10;u=0;a=0.3;Jp=10;JM=0.5*M*a^2;
mu1=a^2*(m+M)+JM; mu2=Jp+m*l^2; mu3=m*a*l;
mug=g*m*l;
for t=0:dt:5,
u=0;
x=x+segway2D_f(x,u)*dt;
segway2D_draw(x);
end;
```

2) The linearization of the system around the operating point $\bar{u} = 0$, $\bar{x} = 0$ is given by (see Exercise 5.8):

$$\dot{x} = \begin{pmatrix} 0 & 0 & 1 & 0 \\ 0 & 0 & 0 & 1 \\ 0 & -\frac{\mu_3 \mu_g}{\mu_1 \mu_2 - \mu_3^2} & 0 & 0 \\ 0 & \frac{\mu_1 \mu_g}{\mu_1 \mu_2 - \mu_3^2} & 0 & 0 \end{pmatrix} x + \begin{pmatrix} 0 \\ 0 \\ \frac{\mu_2 + \mu_3}{\mu_1 \mu_2 - \mu_3^2} \\ -\frac{\mu_1 + \mu_3}{\mu_1 \mu_2 - \mu_3^2} \end{pmatrix} u$$

In order to obtain the state equations of the controller, we add the following function (see function `RegulKLH.m` described on page 133). The main program becomes:

```
x=[0;0;0;0.01]; % initial state of the segway
xr=[0;0;0;0]; % initial state of the observer
dt = 0.01;
m=10;M=1;l=1;g=10;u=0;a=0.3;Jp=10;JM=0.5*M*a^2;
mu1=a^2*(m+M)+JM; mu2=Jp+m*l^2; mu3=m*a*l; mug=g*m*l;
A=[0 0 1 0;0 0 0 1;0 -mu3*mug/(mu1*mu2-mu3^2) 0 0;0
mu1*mug/(mu1*mu2-mu3^2) 0 0]
B=[0;0;(mu2+mu3)/(mu1*mu2-mu3^2);
(-mu1-mu3)/(mu1*mu2-mu3^2)];
```

```
C=[1 0 0 0];
E=[-a 0 0 0]; % since the setpoint variable is xc=-a*x(1)
[Ar,Br,Cr,Dr]=RegulKLH(A,B,C,E,[-2,-2.1,-2.2,-2.3],
[-2,-2.1,-2.2,-2.3]);
w=1; % desired position
for t=0:dt:5;
y=C*x;
u=Cr*xr+Dr*[w;y];
x=x+f(x,u)*dt;
xr=xr+(Ar*xr+Br*[w;y])*dt;
draw_segway(x);
end;
```

3) We evaluate the robustness of the control by adding sensor noise using the `randn()` command, if we need Gaussian noise. For instance, we can add a white Gaussian noise with variance 1 by replacing line `y=C*x` by `y=C*x+randn(size(C*x))`.

4) Following the tank model, we have:

$$\begin{cases} \dot{x} = \dot{\alpha} \cos \psi \\ \dot{y} = \dot{\alpha} \sin \psi \end{cases}$$

Thus, the state equations are:

$$\frac{d}{dt} \begin{pmatrix} \theta \\ \dot{\alpha} \\ \dot{\theta} \\ x \\ y \\ \psi \\ \alpha_1 \\ \alpha_2 \end{pmatrix} = \begin{pmatrix} \dot{\theta} \\ \frac{\mu_3 \left(\mu_2 \dot{\theta}^2 - \mu_g \cos \theta\right) \sin \theta + (\mu_2 + \mu_3 \cos \theta) u_1}{\mu_1 \mu_2 - \mu_3^2 \cos^2 \theta} \\ \frac{\left(\mu_1 \mu_g - \mu_3^2 \dot{\theta}^2 \cos \theta\right) \sin \theta - (\mu_1 + \mu_3 \cos \theta) u_1}{\mu_1 \mu_2 - \mu_3^2 \cos^2 \theta} \\ \cos \psi \cdot \dot{\alpha} \\ \sin \psi \cdot \dot{\alpha} \\ u_2 \\ \dot{\alpha} - u_2 \\ \dot{\alpha} + u_2 \end{pmatrix}$$

with state vector:

$$\mathbf{x} = \left(\theta, \dot{\alpha}, \dot{\theta}, x, y, \psi, \alpha_1, \alpha_2\right)^{\mathrm{T}}$$

5) The MATLAB program that performs this control is composed of the evolution function of the vehicle, given below and of the main program:

```
function dx=segway3d_f(x,u)
theta=x(1); dalpha=x(2); dtheta=x(3); xc=x(4); yc=x(5);
psi=x(6); alpha1=x(7); alpha2=x(8);
c1=cos(x(1));s1=sin(x(1)); den=mu1*mu2-(mu3*c1)^2;
w2=dtheta^2;
ddalpha=(1/den)*(mu3*(mu2*w2-mug*c1)*s1+(mu2+mu3*c1)*u(1));
ddtheta=(1/den)*((mu1*mug-mu3^2*w2*c1)*s1
-(mu1+mu3*c1)*u(1));
dx=[ dtheta; ddalpha; ddtheta; cos(psi)*dalpha; % from
the tank model
sin(psi)*dalpha;   u(2);   dalpha-u(2); dalpha+u(2);];
```

The main program (see file segway3D_main.m) is given below:

```
x=[0;0;0;0;0;0;0;0]; % x=[theta,dalpha,dtheta,x,y,psi,alpha1,alpha2]
dt=0.03; u=[0;0];
m=10;M=1;l=1;g=10;u=0;a=0.3;Jp=10;JM=0.5*M*a^2;
mu1=a^2*(m+M)+JM; mu2=Jp+m*l^2; mu3=m*a*l; mug=g*m*l;
% Construction of the controller which pictures a 2D world
xr2=[0;0;0;0]; % initial state of the observer
A2=[0 0 1 0;0 0 0 1;0 -mu3*mug/(mu1*mu2-mu3^2) 0 0;0
mu1*mug/(mu1*mu2-mu3^2) 0 0];
B2=[0;0; (mu2+mu3)/(mu1*mu2-mu3^2); (-mu1-mu3)/(mu1*mu2-mu3^2)];
C2=[1 0 0 0]; E2=[1 0 0 0];
[Ar2,Br2,Cr2,Dr2]=RegulKLH(A2,B2,C2,E2,[-2,-2.1,-2.2,-2.3],
[-2,-2.1,-2.2,-2.3]);
% The controller is now built
C=[0 0 0 0 0 0 0.5 0.5]; % observation matrix of the system.
```

```
% We measure the average between the angles of the two wheels
w=0; % position setpoint
for t=0:dt:15,
w=w+0.5*dt; % We let the setpoint evolve in order for the segway to
advance.
% Setpoint corresponding to the average angle between the two wheels,
y=C*x+0.05*randn(); % measurement generated by the sensor.
theta=x(1); xc=x(4); yc=x(5); psi=x(6); alpha1=x(7); alpha2=x(8);
psibar=1;u2=-tan(atan(0.5*(psi-psibar))); % rgulation en cap
u=[Cr2*xr2+Dr2*[w;y];u2]; % u1 corresponds to the torque computed by
the 2D controller
theta=x(1); xc=x(4); yc=x(5); psi=x(6); alpha1=x(7); alpha2=x(8);
segway3d_draw(xc,yc,theta,psi,alpha1,alpha2); % We draw the real
segway in 3D.
x=x+segway3d_f(x,u)*dt;
xr2=xr2+(Ar2*xr2+Br2*[w;y])*dt;
end
```

In this program, `xr2` is the state of the pitch controller of the segway. This controller pictures a 2D world.

Solution to Exercise 5.10 (linearization of the tank)

1) We have:

$$\begin{pmatrix} \dot{x} \\ \dot{y} \\ \dot{\theta} \\ \dot{\omega} \\ \dot{v} \end{pmatrix} = \begin{pmatrix} 0 & 0 & -\bar{v}\sin\bar{\theta} & 0 & \cos\bar{\theta} \\ 0 & 0 & \bar{v}\cos\bar{\theta} & 0 & \sin\bar{\theta} \\ 0 & 0 & 0 & 1 & 0 \\ 0 & 0 & 0 & 0 & 0 \\ 0 & 0 & 0 & 0 & 0 \end{pmatrix} \begin{pmatrix} x - \bar{x} \\ y - \bar{y} \\ \theta - \bar{\theta} \\ \omega - \bar{\omega} \\ v - \bar{v} \end{pmatrix} + \begin{pmatrix} 0 & 0 \\ 0 & 0 \\ 0 & 0 \\ 1 & 0 \\ 0 & 1 \end{pmatrix} \begin{pmatrix} u_1 \\ u_2 \end{pmatrix}$$

The state of equilibrium is obtained for $\bar{v} = \bar{\omega} = 0$.

2) The controllability matrix is:

$$\mathcal{C}_{\text{con}} = \begin{pmatrix} 0 & 0 & 0 & \cos\bar\theta & -\bar v \sin\bar\theta & 0 & 0 & 0 & 0 & 0 \\ 0 & 0 & 0 & \sin\bar\theta & \bar v \cos\bar\theta & 0 & 0 & 0 & 0 & 0 \\ 0 & 0 & 1 & 0 & 0 & 0 & 0 & 0 & 0 & 0 \\ 1 & 0 & 0 & 0 & 0 & 0 & 0 & 0 & 0 & 0 \\ 0 & 1 & 0 & 0 & 0 & 0 & 0 & 0 & 0 & 0 \end{pmatrix}$$

which is of rank 5 for $\bar v \neq 0$. At equilibrium, $\bar v \neq 0$ and therefore:

$$\mathcal{C}_{\text{con}} = \begin{pmatrix} 0 & 0 & 0 & \cos\bar\theta & 0 & 0 & 0 & 0 & 0 & 0 \\ 0 & 0 & 0 & \sin\bar\theta & 0 & 0 & 0 & 0 & 0 & 0 \\ 0 & 0 & 1 & 0 & 0 & 0 & 0 & 0 & 0 & 0 \\ 1 & 0 & 0 & 0 & 0 & 0 & 0 & 0 & 0 & 0 \\ 0 & 1 & 0 & 0 & 0 & 0 & 0 & 0 & 0 & 0 \end{pmatrix}$$

which is of rank 4. The system is therefore uncontrollable.

3) The uncontrollable directions **a** are those that are orthogonal to all the columns of $\mathcal{C}_{\text{con}}$. In order to find them, we need to solve:

$$\mathcal{C}_{\text{con}}^{\text{T}} \cdot \mathbf{a} = \mathbf{0}, \text{ with } \mathbf{a} \neq \mathbf{0}$$

and thus, for $\bar v = 0$, we obtain:

$$\mathbf{a} = \lambda.(\sin\theta,\ -\cos\theta,\ 0,\ 0\ 0)^{\text{T}} \text{ with } \lambda \neq 0$$

Therefore we cannot move perpendicularly to the direction of the tank (in first order). In second order, it is possible ("slot" principle when we park a car).

Solution to Exercise 5.11 (linearization of the hovercraft)

1) By linearizing, we obtain the following approximation:

$$\dot{\mathbf{x}} \simeq \mathbf{f}(\bar{\mathbf{x}}, \bar{\mathbf{u}}) + \frac{\partial \mathbf{f}}{\partial \mathbf{x}}(\bar{\mathbf{x}}, \bar{\mathbf{u}}) \cdot (\mathbf{x} - \bar{\mathbf{x}}) + \frac{\partial \mathbf{f}}{\partial \mathbf{u}}(\bar{\mathbf{x}}, \bar{\mathbf{u}}) \cdot (\mathbf{u} - \bar{\mathbf{u}})$$

i.e.:

$$\begin{pmatrix} \dot{x} \\ \dot{y} \\ \dot{\theta} \\ \dot{v}_x \\ \dot{v}_y \\ \dot{\omega} \end{pmatrix} \simeq \begin{pmatrix} \bar{v}_x \\ \bar{v}_y \\ \bar{\omega} \\ \bar{u}_1 \cos \bar{\theta} \\ \bar{u}_1 \sin \bar{\theta} \\ \bar{u}_2 \end{pmatrix} + \begin{pmatrix} 0 & 0 & 0 & 1 & 0 & 0 \\ 0 & 0 & 0 & 0 & 1 & 0 \\ 0 & 0 & 0 & 0 & 0 & 1 \\ 0 & 0 & -\bar{u}_1 \sin \bar{\theta} & 0 & 0 & 0 \\ 0 & 0 & \bar{u}_1 \cos \bar{\theta} & 0 & 0 & 0 \\ 0 & 0 & 0 & 0 & 0 & 0 \end{pmatrix} \begin{pmatrix} x - \bar{x} \\ y - \bar{y} \\ \theta - \bar{\theta} \\ v_x - \bar{v}_x \\ v_y - \bar{v}_y \\ \omega - \bar{\omega} \end{pmatrix}$$

$$+ \begin{pmatrix} 0 & 0 \\ 0 & 0 \\ 0 & 0 \\ \cos \bar{\theta} & 0 \\ \sin \bar{\theta} & 0 \\ 0 & 1 \end{pmatrix} \begin{pmatrix} u_1 - \bar{u}_1 \\ u_2 - \bar{u}_2 \end{pmatrix}$$

2) The controllability matrix is:

$$\mathcal{C}_{\text{con}} = \begin{pmatrix} 0 & 0 & \cos\bar{\theta} & 0 & 0 & 0 & 0 & -\bar{u}_1 \sin\bar{\theta} & 0 & 0 & 0 & 0 \\ 0 & 0 & \sin\bar{\theta} & 0 & 0 & 0 & 0 & \bar{u}_1 \cos\bar{\theta} & 0 & 0 & 0 & 0 \\ 0 & 0 & 0 & 1 & 0 & 0 & 0 & 0 & 0 & 0 & 0 & 0 \\ \cos\bar{\theta} & 0 & 0 & 0 & 0 & -\bar{u}_1 \sin\bar{\theta} & 0 & 0 & 0 & 0 & 0 & 0 \\ \sin\bar{\theta} & 0 & 0 & 0 & 0 & \bar{u}_1 \cos\bar{\theta} & 0 & 0 & 0 & 0 & 0 & 0 \\ 0 & 1 & 0 & 0 & 0 & 0 & 0 & 0 & 0 & 0 & 0 & 0 \end{pmatrix}$$

3) This matrix is not of full rank if:

$$\det \begin{pmatrix} 0 & \cos\bar{\theta} & 0 & -\bar{u}_1 \sin\bar{\theta} \\ 0 & \sin\bar{\theta} & 0 & \bar{u}_1 \cos\bar{\theta} \\ \cos\bar{\theta} & 0 & -\bar{u}_1 \sin\bar{\theta} & 0 \\ \sin\bar{\theta} & 0 & \bar{u}_1 \cos\bar{\theta} & 0 \end{pmatrix} = 0$$

in other words if:

$$\bar{u}_1^2 \cdot \left(\cos^2 \bar{\theta} + \sin^2 \bar{\theta}\right)^2 = 0$$

Thus, if $\bar{u}_1 = 0$, the matrix is not of full rank.

4) In order to find the non-controllable directions in the case where $u_1 = 0$, we need to find a vector orthogonal to all the columns of the controllability matrix. We solve:

$$\mathcal{C}_{\text{con}}^{\text{T}} \cdot \mathbf{a} = \mathbf{0}, \text{ with } \mathbf{a} \neq \mathbf{0}$$

We obtain:

$$\mathbf{a} = \begin{pmatrix} x \\ y \\ \theta \\ v_x \\ v_y \\ \omega \end{pmatrix} = \begin{pmatrix} -\alpha \sin \bar{\theta} \\ \alpha \cos \bar{\theta} \\ 0 \\ -\beta \sin \bar{\theta} \\ \beta \cos \bar{\theta} \\ 0 \end{pmatrix}$$

where α and β are constants. We cannot therefore control the lateral skid (normal to the axis of the hovercraft) if we do not accelerate.

Solution to Exercise 5.12 (linearization of the single-acting cylinder)

1) The condition $\dot{\mathbf{x}} = \mathbf{0}$ gives:

$$\begin{cases} x_2 = 0 \\ ax_3 - kx_1 = 0 \\ x_3 \left(x_2 - \frac{u}{a}\right) = 0 \end{cases}$$

in other words:

$$\begin{cases} x_2 = 0 \\ ax_3 - kx_1 = 0 \\ u = 0 \end{cases}$$

The operating points are therefore to the form:

$$(\bar{\mathbf{x}}, \bar{u}) = \left(\bar{x}_1, 0, \frac{k}{a}\bar{x}_1, 0\right)$$

2) Let us now differentiate the system around the operating point $(\bar{\mathbf{x}}, \bar{u})$ in order to obtain an affine approximation of it. We obtain the linearized system:

$$\dot{\mathbf{x}} = \begin{pmatrix} 0 & 1 & 0 \\ -\frac{k}{m} & 0 & \frac{a}{m} \\ 0 & -\frac{k}{a} & 0 \end{pmatrix}(\mathbf{x} - \bar{\mathbf{x}}) + \begin{pmatrix} 0 \\ 0 \\ \frac{k}{a^2} \end{pmatrix} u$$

Note that the coefficients of the matrices **A** and **B** do not depend on $\bar{\mathbf{x}}$. This is quite rare, and it means that the behavior of a linear control will not depend on the chosen operating point. Let us note that the absence of the matrices **C** and **D** in the linearized system is a consequence of the fact that the nonlinear system considered is autonomous (i.e. without output).

Solution to Exercise 5.13 (pole placement control of a nonlinear system)

1) At equilibrium, we have to have $f(\bar{x}, \bar{u}) = 0$. For $\bar{x} = 2$, we need to have $\bar{u} = -8$. If we want that at equilibrium $y = w$, we will need to take $E = 3$. If, moreover, we would like all the poles of the looped system to be equal to -1, we would need $p_{\text{con}} = p_{\text{obs}} = -1$. The linearization around $(\bar{x}, \bar{u})$ gives us $A = 8$, $B = 1$ and $C = 3$. For K and L, we need to solve the two polynomial equations:

$$\begin{cases} \det(sI - A + BK) = s + 1 \\ \det(sI - A^{\mathrm{T}} + C^{\mathrm{T}}L^{\mathrm{T}}) = s + 1 \end{cases}$$

i.e. $s-8+K = s+1$ and $s-8+3L = s+1$. Therefore, $K = 9$ and $L = 3$. Moreover, $\bar{y} = 6$, $\bar{w} = 3\bar{x} = 6$. $H = -\left(3(8-9)^{-1}\right)^{-1} = 1/3$. Therefore, the controller is written as:

$$\mathcal{R} : \begin{cases} \frac{d}{dt}\hat{x} = -10\hat{x} + (w-6)/3 + 3(y-6) \\ u = -8 - 9\hat{x} + (w-6)/3 \end{cases}$$

2) The state equations of our looped system are:

$$\begin{cases} \dot{x} = 2x^2 - 8 - 9\hat{x} + \frac{w-6}{3} \\ \frac{d}{dt}\hat{x} = -10\hat{x} + \frac{w-6}{3} + 3(3x-6) \\ y = 3x \end{cases}$$

Solution to Exercise 5.14 (state feedback of a wiring system)

1) We have:

$$\begin{pmatrix} \dot{x}_1 \\ \dot{x}_2 \end{pmatrix} = \begin{pmatrix} x_2 \\ u - \sin x_1 - x_2 \end{pmatrix}$$

These equations are those of a damped pendulum with angle $\theta = x_1$ and angular velocity $\dot{\theta} = x_2$.

2) We solve $\mathbf{f}(\bar{\mathbf{x}}, \bar{u}) = \mathbf{0}$ with $\bar{u} = 0$. We find:

$$\bar{\mathbf{x}} = \begin{pmatrix} k\pi \\ 0 \end{pmatrix} \text{ with } k \in \mathbb{Z}$$

3) We take $\bar{\mathbf{x}} = (\pi, 0)^T$. Therefore:

$$\begin{pmatrix} \dot{x}_1 \\ \dot{x}_2 \end{pmatrix} \simeq \mathbf{f}(\bar{\mathbf{x}}, \bar{u}) + \begin{pmatrix} 0 & 1 \\ -\cos \bar{x}_1 & -1 \end{pmatrix} (\mathbf{x} - \bar{\mathbf{x}}) + \begin{pmatrix} 0 \\ 1 \end{pmatrix} u$$

$$= \underbrace{\begin{pmatrix} 0 & 1 \\ 1 & -1 \end{pmatrix}}_{\mathbf{A}} \cdot \underbrace{\begin{pmatrix} x_1 - \pi \\ x_2 \end{pmatrix}}_{\mathbf{x} - \bar{\mathbf{x}}} + \underbrace{\begin{pmatrix} 0 \\ 1 \end{pmatrix}}_{\mathbf{B}} u$$

The characteristic polynomial is $P(s) = s^2 + s - 1$. Since $P(0) = -1$ and that P has a positive curvature, it necessarily has a positive real root. The system is therefore unstable.

4) To calculate **K**, we solve:

$$\det(sI - \mathbf{A} + \mathbf{BK}) = (s+1)^2.$$

i.e.:

$$\det\left(\begin{pmatrix} s & 0 \\ 0 & s \end{pmatrix} - \begin{pmatrix} 0 & 1 \\ 1 & -1 \end{pmatrix} + \begin{pmatrix} 0 \\ 1 \end{pmatrix}\begin{pmatrix} k_1 & k_2 \end{pmatrix}\right) = \det\begin{pmatrix} s & -1 \\ k_1 - 1 & s + k_2 + 1 \end{pmatrix}$$
$$= s^2 + s(k_2 + 1) + k_1 - 1 \qquad\qquad\qquad\qquad\qquad = s^2 + 2s + 1$$

By identification, we obtain $k_1 = 2$, $k_2 = 1$. The controller is therefore:

$$u = (-2,\ -1)(\mathbf{x} - \bar{\mathbf{x}}) = -2x_1 + 2\pi - x_2$$

Solution to Exercise 5.15 (controlling an inverted rod pendulum in MATLAB)

1) In order to linearize this system, we first of all need to calculate an operating point. For this, we solve $\mathbf{f}(\bar{\mathbf{x}}, \bar{\mathbf{u}}) = \mathbf{0}$, in other words:

$$\begin{cases} \bar{x}_3 = 0 \\ \bar{x}_4 = 0 \\ m\sin\bar{x}_2(g\cos\bar{x}_2 - \ell\bar{x}_4^2) + \bar{u} = 0 \\ \sin\bar{x}_2((M+m)g - m\ell\bar{x}_4^2\cos\bar{x}_2) + \cos\bar{x}_2\bar{u} = 0 \end{cases}$$

Since $\bar{x}_4 = 0$, the last two lines can be written in the form:

$$\begin{pmatrix} mg\cos\bar{x}_2 & 1 \\ (M+m)g & \cos\bar{x}_2 \end{pmatrix}\begin{pmatrix} \sin\bar{x}_2 \\ \bar{u} \end{pmatrix} = \begin{pmatrix} 0 \\ 0 \end{pmatrix}$$

The determinant of the matrix, however, is zero if:

$$mg\cos^2\bar{x}_2 - (M+m)g = 0$$

in other words:

$$\cos^2 \bar{x}_2 = \frac{M+m}{m}$$

which is impossible since $\frac{M+m}{m} > 1$. Form this, we deduce that $\sin \bar{x}_2 = 0$, $\bar{u} = 0$. The operating points are of the form:

$$\bar{\mathbf{x}} = (\bar{x}_1, k\pi, 0, 0)^{\mathrm{T}} \text{ and } \bar{u} = 0 \text{ with } k \in \mathbb{Z}$$

2) Around the operating point $\bar{\mathbf{x}} = (0,0,0,0)$ and $\bar{u} = 0$, let us linearize the system by applying the limited development method. We have:

$$\begin{cases} \frac{m\sin x_2(g\cos x_2 - \ell x_4^2) + u}{M + m\sin^2 x_2} = \frac{m(x_2+\varepsilon)(g(1+\varepsilon) - \ell\varepsilon) + u}{M + m(x_2+\varepsilon)^2} \\ = \frac{m(x_2+\varepsilon)(g+\varepsilon) + u}{M+\varepsilon} = \frac{mgx_2 + u}{M} + \varepsilon \\ \frac{\sin x_2((M+m)g - m\ell x_4^2 \cos x_2) + \cos x_2 u}{\ell(M + m\sin^2 x_2)} \\ = \frac{(x_2+\varepsilon)((M+m)g - m\ell\varepsilon(1+\varepsilon)) + (1+\varepsilon)u}{\ell(M + m(x_2+\varepsilon)^2)} \\ = \frac{x_2(M+m)g + u}{\ell M} + \varepsilon \end{cases}$$

Thus, we obtain the linearized system:

$$\begin{cases} \dot{\mathbf{x}} = \begin{pmatrix} 0 & 0 & 1 & 0 \\ 0 & 0 & 0 & 1 \\ 0 & \frac{mg}{M} & 0 & 0 \\ 0 & \frac{(M+m)g}{M\ell} & 0 & 0 \end{pmatrix} \mathbf{x} + \begin{pmatrix} 0 \\ 0 \\ \frac{1}{M} \\ \frac{1}{M\ell} \end{pmatrix} u \\ y = \begin{pmatrix} 1 & 0 & 0 & 0 \end{pmatrix} \mathbf{x} \end{cases}$$

Here, since $\bar{\mathbf{x}} = 0$, the quantities $\tilde{\mathbf{x}}$ and $\mathbf{x}$ are the same and it is for this reason that we have kept the notation $\mathbf{x}$ in the above state equation. It is easy to verify that the linearized system of our inverted rod pendulum is observable and controllable for the nominal values of the parameters.

3) The REGULKLH algorithm (refer to function RegulKLH.m described on page 133) allows us to obtain our controller. The MATLAB instructions are the following:

```
A=[0 0 1 0;0 0 0 1;0 m*g/M 0 0;0 (M+m)*g/(l*M)
 0 0];
B=[0;0;1/M;1/(l*M)];
C=[1 0 0 0];
E=[1 0 0 0]; % setpoint matrix
pcom=[-2 -2.1 -2.2 -2.3];
pobs=[-2 -2.1 -2.2 -2.3]);
K=place(A,B,pcom);
L=place(A',C',pobs)';
H=-inv(E*inv(A-B*K)*B);
Ar=A-B*K-L*C;Br=[B*H L];
Cr=-K;Dr=[H,zeros(size(B'*C'))];
```

In this program, `Ar`, `Br`, `Cr`, `Dr` are the state matrices for the controller. Note that due to the algorithm used by MATLAB, the `place` command requests distinct poles. The entire MATLAB program can be found in `pendinv_main.m`.

4) The mechanical energy is given by:

$$E_m(\mathbf{x}) = \underbrace{\frac{1}{2}m\ell^2\dot{\theta}^2}_{\text{kinetic energy}} + \underbrace{mg\ell(\cos\theta - 1)}_{\text{potential energy}}$$

$$= \frac{1}{2}m\ell^2 x_4^2 + mg\ell(\cos x_2 - 1)$$

The potential energy constant has been chosen to be zero in the unstable point of equilibrium of the pendulum. Let us take $V(\mathbf{x}) = E_m^2(\mathbf{x})$. This function will be minimal when the pendulum is in upper equilibrium. We will therefore try to minimize $V(\mathbf{x})$. We have:

$$\dot{V}(\mathbf{x}) = \frac{dV}{d\mathbf{x}}(\mathbf{x}).\dot{\mathbf{x}} = 2E_m(\mathbf{x}).\frac{dE_m}{d\mathbf{x}}(\mathbf{x}).\mathbf{f}(\mathbf{x},\mathbf{u})$$

$$= m\left(\ell^2 x_4^2 + 2g\ell(\cos x_2 - 1)\right)\begin{pmatrix} 0 & -mg\ell(\sin x_2) & 0 & m\ell^2 x_4 \end{pmatrix}$$

$$\cdot\left(\begin{pmatrix} x_3 \\ x_4 \\ \frac{-m(\sin x_2)(\ell x_4^2 - g\cos x_2)}{M + m\sin^2 x_2} \\ \frac{(\sin x_2)((M+m)g - m\ell x_4^2 \cos x_2)}{\ell(M + m\sin^2 x_2)} \end{pmatrix} + \begin{pmatrix} 0 \\ 0 \\ \frac{1}{M + m\sin^2 x_2} \\ \frac{(\cos x_2)}{\ell(M + m\sin^2 x_2)} \end{pmatrix} u\right)$$

$$= a(\mathbf{x}).u + b(\mathbf{x})$$

To raise the pendulum, we will need to take:

$$u = -u_{\max}.\text{sign}(a(\mathbf{x}))$$

$$= -u_{\max}.\text{sign}\left(\left(\ell^2 x_4^2 + 2g\ell(\cos x_2 - 1)\right)\frac{m\ell x_4 \cos x_2}{M + m\sin^2 x_2}\right)$$

$$= -u_{\max}.\text{sign}\left(\left(\ell x_4^2 + 2g(\cos x_2 - 1)\right)(x_4 \cos x_2)\right)$$

The corresponding program, which can be found in pendinv_monte.m, is given below.

```
m=1;M=5;l=1;g=9.81; dt=0.01;
x=[0;3;0.4;0]; % bottom-positioned pendulum
for t=0:dt:12,
u=-5*sign(l*x(4)^2+2*g*(cos(x(2))-1)*x(4)
*cos(x(2)));
x=x+pendinv_f(x,u)*dt;
end
```

Solution to Exercise 5.16 (controlling a car along a wall in MATLAB)

1) In order to understand the following proof, it is important to recall the meaning of the sign of the determinant of the two

vectors $\vec{u}$ and $\vec{v}$ of $\mathbb{R}^2$. We have (i) $\det(\vec{u}, \vec{v}) > 0$ if $\vec{v}$ is on the left-hand side of $\vec{u}$, (ii) $\det(\vec{u}, \vec{v}) < 0$ if $\vec{v}$ is on the right-hand side of $\vec{u}$ and (iii) $\det(\vec{u}, \vec{v}) = 0$ if $\vec{u}$ and $\vec{v}$ are collinear. Thus, for instance, on the left-hand side of Figure 5.10, $\det(a - m, \vec{u}) > 0$ and $\det(b - m, \vec{u}) < 0$. Let us recall as well that the determinant is a multi-linear form, in other words:

$$\begin{aligned}\det(a\mathbf{u} + b\mathbf{v}, c\mathbf{x} + d\mathbf{y}) &= a\det(\mathbf{u}, c\mathbf{x} + d\mathbf{y}) + b\det(\mathbf{v}, c\mathbf{x} + d\mathbf{y}) \\ &= ac\det(\mathbf{u}, \mathbf{x}) + bc\det(\mathbf{v}, \mathbf{x}) \\ &\quad + ad\det(\mathbf{u}, \mathbf{y}) + bd\det(\mathbf{v}, \mathbf{y})\end{aligned}$$

The line $\mathcal{D}(m, \vec{u})$ that passes through the point m and the direction vector $\vec{u}$ cuts the plane into two half-planes: those that satisfy $\det(z - m, \vec{u}) \geq 0$ and those that satisfy $\det(z - m, \vec{u}) \leq 0$. It cuts therefore the segment [ab] if a and b are in different half-planes, in other words, if $\det(a - m, \vec{u}) \cdot \det(b - m, \vec{u}) \leq 0$.

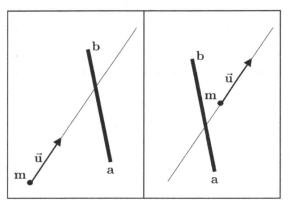

Figure 5.10. *The line $\mathcal{D}(m, \tilde{u})$ cuts the segment* [ab]

In the situation of Figure 5.10 on the left-hand side, the half-line $\mathcal{E}(m, \vec{u})$ with vertex m and direction vector $\vec{u}$ cuts the segment [ab]. This is not the case on the right sub-figure. Our condition is therefore not sufficient in order to state that $\mathcal{E}(m, \vec{u})$ will cut [ab]. Let us assume that

$\det(\mathbf{a} - \mathbf{m}, \vec{\mathbf{u}}) \cdot \det(\mathbf{b} - \mathbf{m}, \vec{\mathbf{u}}) \leq 0$, in other words, that the line $\mathcal{D}(\mathbf{m}, \vec{\mathbf{u}})$ cuts the segment $[\mathbf{ab}]$. The points of the half-line $\mathcal{E}(\mathbf{m}, \vec{\mathbf{u}})$ all satisfy $\mathbf{z} = \mathbf{m} + \alpha \vec{\mathbf{u}}$, $\alpha \geq 0$. As illustrated in Figure 5.11, $\mathbf{m}$ is on the segment $[\mathbf{ab}]$ when the vectors $\mathbf{m} + \alpha \vec{\mathbf{u}} - \mathbf{a}$ and $\mathbf{b} - \mathbf{a}$ are collinear (see figure), in other words, when α satisfies $\det(\mathbf{m} + \alpha \vec{\mathbf{u}} - \mathbf{a}, \mathbf{b} - \mathbf{a}) = 0$.

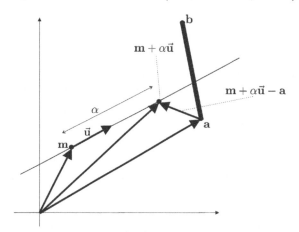

Figure 5.11. *The point* $\mathbf{m}$ *is on the segment* $[\mathbf{ab}]$ *when the vectors* $\mathbf{m} + \alpha \vec{\mathbf{u}} - \mathbf{a}$ *and* $\mathbf{b} - \mathbf{a}$ *are collinear*

Given the multi-linearity of the determinant, this equation can be re-written as:

$$\det(\mathbf{m} - \mathbf{a}, \mathbf{b} - \mathbf{a}) + \alpha \det(\vec{\mathbf{u}}, \mathbf{b} - \mathbf{a}) = 0$$

By isolating α, we obtain:

$$\alpha = \frac{\det(\mathbf{a} - \mathbf{m}, \mathbf{b} - \mathbf{a})}{\det(\vec{\mathbf{u}}, \mathbf{b} - \mathbf{a})}$$

If $\alpha \geq 0$, then α represents the distance d traveled by the ray starting from $\mathbf{m}$ in the direction $\vec{\mathbf{u}}$ before meeting the segment. If $\alpha < 0$, this means that the ray will never meet the segment since the latter is on the wrong side.

2) The evolution function is given below.

```
function xdot=f(x,u)
xdot=[x(4)*cos(x(5))*cos(x(3));
x(4)*cos(x(5))*sin(x(3));
x(4)*sin(x(5))/3;
u(1);u(2)];
end
```

The observation function of our system given below is a direct consequence of the first question. This function returns the distance d measured by the rangefinder, the velocity v of the font wheels and the angle of the steering wheel δ. The function first of all calculates the vector u that indicates the direction of the laser rangefinder and the point m where the laser leaves from. The distance d returned must be the smallest one among all the distances likely to be returned by each of the segments. In the program, function P represents the matrix containing the vertices of the polygon. Finally, let us note that this function will not be used by the controller, but only in order to simulate our real system. Indeed, our controller does not know the shape of the polygon around which the car is supposed to be turning: it thinks that the car is driving along an infinite straight wall.

```
function y=circuit_g(x,P)
u=[sin(x(3));-cos(x(3))]; m=[x(1);x(2)]; d=Inf;
for j=1:size(P,2)-1
a=P(:,j); b=P(:,j+1);
if ((det([a-m u])*det([b-m u]) <= 0)&&(det([a-m b-a])/det([u b-a])>=0))
d=min(det([a-m b-a])/det([u b-a]),d);
end
end
y=[d;x(4);x(5)];
end
```

3) For the evolution equation of our system as seen by the controller, we will use the reasoning seen in Exercise 1.11 to model the car, except that y no longer exists. One must be careful with the sign $-$ in the equation $\dot{x} = -v\cos\delta\cos\theta$ since in our new model, x increases when the car is moving toward the left. We thus have:

$$\begin{pmatrix} \dot{x} \\ \dot{\theta} \\ \dot{v} \\ \dot{\delta} \end{pmatrix} = \begin{pmatrix} -v\cos\delta\cos\theta \\ \frac{v\sin\delta}{L} \\ u_1 \\ u_2 \end{pmatrix}$$

For the observation equation, we have:

$$\mathbf{y} = \begin{pmatrix} \frac{x}{\sin\theta} \\ v \\ \delta \end{pmatrix} \begin{array}{l} \rightarrow \text{ distance returned by the rangefinder} \\ \rightarrow \text{ velocity of the front wheels} \\ \rightarrow \text{ angle of the steering wheel} \end{array}$$

where L is the distance between the front and rear axles.

4) The linearization around the operating point and for $L = 3$ gives us the matrices of the linearized system:

$$\mathbf{A} = \begin{pmatrix} 0 & 7 & 0 & 0 \\ 0 & 0 & 0 & \frac{7}{3} \\ 0 & 0 & 0 & 0 \\ 0 & 0 & 0 & 0 \end{pmatrix}, \mathbf{B} = \begin{pmatrix} 0 & 0 \\ 0 & 0 \\ 1 & 0 \\ 0 & 1 \end{pmatrix},$$

$$\mathbf{C} = \begin{pmatrix} 1 & 0 & 0 & 0 \\ 0 & 0 & 1 & 0 \\ 0 & 0 & 0 & 1 \end{pmatrix}, \mathbf{D} = \begin{pmatrix} 0 & 0 \\ 0 & 0 \\ 0 & 0 \end{pmatrix}$$

5) The set point matrix is given by:

$$\mathbf{E} = \begin{pmatrix} 1 & 0 & 0 & 0 \\ 0 & 0 & 1 & 0 \end{pmatrix}$$

By choosing as poles $p_{con} = p_{obs} = (-2, -2, -2, -2)$, the REGULKLH algorithm $(A, B, C, E, p_{con}, p_{obs})$ generated for us the following controller:

$$\begin{cases} \frac{d}{dt}\hat{x} = \begin{pmatrix} -4 & 7 & 0 & 0 \\ -0.6 & 0 & 0 & 0 \\ 0 & 0 & -4 & 0 \\ -0.5 & -5.1 & 0 & -8 \end{pmatrix} \hat{x} + \begin{pmatrix} 0 & 0 & 4 & 0 & 0 \\ 0 & 0 & 0.6 & 0 & 2.3 \\ 0 & 2 & 0 & 2 & 0 \\ 0.5 & 0 & 0 & 0 & 2 \end{pmatrix} \begin{pmatrix} w \\ y \end{pmatrix} \\ \qquad - \begin{pmatrix} 20 \\ 2.9 \\ 28 \\ 2.4 \end{pmatrix} \\ u = \begin{pmatrix} 0 & 0 & -2 & 0 \\ -0.49 & -5.14 & 0 & -6 \end{pmatrix} \hat{x} + \begin{pmatrix} 0 & 2 \\ 0.48 & 0 \end{pmatrix} \left(w - \begin{pmatrix} 5 \\ 7 \end{pmatrix} \right) \end{cases}$$

6) The corresponding program is given below (refer to the file `circuit_main.m`).

```
r0=5; v0=7; dt=0.05;
P=[-10 -10 0 10 20 32 35 30 20 0 -10 ; -5 5 15 20 20 15 10 0
-3 -6 -5];
A=[0 7 0 0;0 0 0 7/3;0 0 0 0;0 0 0 0]; B=[0 0;0 0;1 0;0 1];
C=[1 0 0 0;0 0 1 0;0 0 0 1]; E=[1 0 0 0;0 0 1 0];
[Ar,Br,Cr,Dr]=RegulKLH(A,B,C,E,[-2 -2.1 -2.2 -2.3],[-2 -2.1
-2.2 -2.3]);
ubar=[0;0]; xbar=[r0;1.57;v0;0];
wbar=E*xbar; ybar=[xbar(1)/sin(xbar(2));xbar(3);xbar(4)];
x=[-15;0;pi/2;7;0.1]; xr=[0;0;0;0];
for t=0:dt:40,
w=[r0;v0]; y=circuit_g(x,P); u=ubar+Cr*xr+Dr*[w-wbar;y-ybar];
x1=x+circuit_f(x,u)*dt; xr1=xr+(Ar*xr+Br*[w-wbar;y-ybar])*dt;
x=x1;xr=xr1;
end;
```

Solution to Exercise 5.17 (nonlinear control)

First part

1) The state representation is given by:

$$\dot{\mathbf{x}} = \begin{pmatrix} 0 & 1 \\ 0 & 0 \end{pmatrix} \mathbf{x} + \begin{pmatrix} 0 \\ 1 \end{pmatrix} u$$

$$y = \begin{pmatrix} 1 & 0 \end{pmatrix} \mathbf{x}$$

2) We have $\ddot{y} = k_1(w-y) + k_2(\dot{w}-\dot{y}) + \ddot{w}$. Thus, if $e = w-y$, we obtain:

$$\ddot{e} + k_1 \dot{e} + k_2 e = 0$$

3) We have:

$$s^2 + k_1 s + k_2 = (s+1)^2$$

or, by identification $k_1 = 1, k_2 = 2$.

Second part

1) We have:

$$\begin{cases} \ddot{x} = \dot{v}\cos\theta - \dot{\theta}v\sin\theta = u_2\cos\theta - u_1 v\sin\theta \\ \ddot{y} = \dot{v}\sin\theta + \dot{\theta}v\cos\theta = u_2\sin\theta + u_1 v\cos\theta \end{cases}$$

Thus:

$$\begin{pmatrix} \ddot{x} \\ \ddot{y} \end{pmatrix} = \underbrace{\begin{pmatrix} -v\sin\theta & \cos\theta \\ v\cos\theta & \sin\theta \end{pmatrix}}_{\mathbf{A}(\mathbf{x})} \begin{pmatrix} u_1 \\ u_2 \end{pmatrix}$$

2) By taking $\mathbf{u} = \mathbf{A}^{-1}(\mathbf{x})\mathbf{q}$, we obtain:

$$\begin{pmatrix} \ddot{x} \\ \ddot{y} \end{pmatrix} = \begin{pmatrix} q_1 \\ q_2 \end{pmatrix}$$

in other words, a system composed of two decoupled integrators. A proportional and derivative-type control on each of these chains gives us:

$$\begin{pmatrix} q_1 \\ q_2 \end{pmatrix} = \begin{pmatrix} (x_a - x) + 2(\dot{x}_a - \dot{x}) + \ddot{x}_a \\ (y_a - y) + 2(\dot{y}_a - \dot{y}) + \ddot{y}_a \end{pmatrix}$$

However, $\dot{x} = v \cos \theta$ and $\dot{y} = v \sin \theta$. The resulting control is therefore:

$$\begin{pmatrix} u_1 \\ u_2 \end{pmatrix} = \begin{pmatrix} -v \sin \theta & \cos \theta \\ v \cos \theta & \sin \theta \end{pmatrix}^{-1} \begin{pmatrix} (x_a - x) + 2(\dot{x}_a - v \cos \theta) + \ddot{x}_a \\ (y_a - y) + 2(\dot{y}_a - v \sin \theta) + \ddot{y}_a \end{pmatrix}$$

By applying this control rule, we have the guarantee that the error will converge toward zero at e^{-t}.

Bibliography

[BAC 92] BACCELLI F., COHEN G., OLSDER G.J., et al., *Synchronization and Linearity: An Algebra for Discrete Event Systems*, John Wiley & Sons, New York, 1992.

[BOU 06] BOURLÈS H., *Systèmes linéaires; de la modélisation à la commande*, Hermès, Paris, 2006.

[BOY 06] BOYER F., POREZ M., KHALIL W., "Macro-continuous computed torque algorithm for a three-dimensional eel-like robot", *IEEE Transactions on Robotics*, vol. 22, no. 4, pp. 763–775, 2006.

[JAU 04] JAULIN L., "Modélisation et commande d'un bateau à voile", *CIFA (Conférence Internationale Francophone d'Automatique)*, CDROM, Douz, Tunisie, 2004.

[JAU 05] JAULIN L., *Représentation d'état pour la modélisation et la commande des systèmes*, Hermès, Paris, 2005.

[JAU 09] JAULIN L., "Robust set membership state estimation: application to underwater robotics", *Automatica*, vol. 45, no. 1, pp. 202–206, 2009.

[JAU 10] JAULIN L., Commande d'un skate-car par biomimétisme, *CIFA*, Nancy, France, 2010.

[JAU 12a] JAULIN L., LE BARS F., "An interval approach for stability analysis: application to sailboat robotics", *IEEE Transaction on Robotics*, vol. 27, no. 5, 2012.

[JAU 12b] JAULIN L., LE BARS F., "A simple controller for line following of sailboats", *5th International Robotic Sailing Conference*, Springer, Cardiff, Wales, England, pp. 107–119, 2012.

[KAI 80] KAILATH T., *Linear Systems*, Prentice Hall, Englewood Cliffs, 1980.

[KHA 02] KHALIL H.K., *Nonlinear Systems*, Prentice Hall, 2002.

[KHA 07] KHALIL W., DOMBRE E., *Robot Manipulators: Modeling, Performance Analysis and Control*, ISTE, London and John Wiley & Sons, New York, 2007.

[LAU 01] LAUMOND J.P., *La Robotique Mobile*, Hermès, Paris, 2001.

[RIV 89] RIVOIRE M., FERRIER J.L., *Cours et exercices d'automatique, Tomes 1, 2 and 3*, Eyrolles, Paris, 1989.

[WAL 14] WALTER E., *Numerical Methods and Optimization: A Consumer Guide*, Springer, London, 2014.

Index

C

change of basis, 97, 98, 111–114, 138, 162
control, 127, 185
controllability, 127, 128, 135–137, 151–153, 155, 174, 195–198, 212, 213, 219–221
controller, 28, 32, 67, 129, 130, 131, 133, 140, 142–150, 160, 167, 171, 173, 180, 183, 189, 190, 196, 199, 200–206, 215, 217, 218, 223–226, 331, 332

D, E, G

dynamics, 2, 3, 6, 18, 21–25, 28, 29, 36, 37, 43, 61, 131, 132
Euler's method, 54–56, 58, 60, 61, 66, 68, 69, 73, 82, 196, 202
graphics, 47

H, I, J

homogeneous coordinates, 52, 59, 62, 63, 70, 76
hydraulic system, 17
integration method, 54, 56
inverted rod pendulum, 6–8, 24, 26–28, 200, 224, 225
Jordan normal form, 97, 102, 112, 125

K, L

Kalman decomposition, 138, 139, 158
Laplace transform, 87–91, 93, 108
linear system, 4, 16, 40, 85–88, 92, 93, 95, 97, 100–102, 110, 112, 121, 124, 128, 129, 137, 138, 141, 144, 152, 159, 185, 188, 189, 212
linearization, 185, 187, 191–194, 197, 198, 208, 209, 210, 211, 215, 218, 219, 221, 222, 231
linearized control, 185

loop, 8, 11, 32, 60, 67, 73, 94,
    127, 134, 139, 145, 167, 207
Luenberger, 131, 176

M

mechanical systems, 3, 33, 73
mobile robots, 3
modal form, 102, 123, 124
modeling, 1, 3, 6–8, 11, 12, 24,
    26, 32, 34, 36, 149

O

observability, 127–129, 135,
    137, 138, 151, 152, 156,
    157, 175
observer, 34, 131, 146–148,
    163, 173–177, 182
operating point, 2, 127, 187,
    188, 190, 196–199, 201,
    204, 212, 213, 215, 222,
    224, 225, 231
output feedback controller,
    131, 133, 140, 148–150,
    196, 201

P, R

path, 58, 60, 68, 69, 73, 138,
    206
point of equilibrium, 5, 57, 60,
    68, 72, 141, 187, 193, 194,
    196, 200, 210, 226
polarization, 187, 188
pole placement, 128, 130–132,
    139, 140, 147, 158, 159,
    173, 181, 196, 199, 205, 222
Runge-Kutta method, 55, 56,
    60

S

segway, 8, 28, 195–197, 212,
    214, 215–218

servomotor, 4, 7, 8
simulation, 36, 47, 49, 51,
    53–55, 59, 60, 62, 65, 67,
    71, 74, 76, 78, 80, 83, 202
skating robot, 65–67
stability, 11, 73, 85, 86, 87, 92,
    104, 106, 146, 195, 211
state
    feedback controller, 143,
        146, 173, 200
    representation, 1, 4, 21, 22,
        40, 87, 91, 95, 100, 112,
        121, 123, 140, 142, 146,
        148, 160, 165, 179, 205,
        233
systems, 1–3, 9, 33, 47, 49, 54,
    68, 73, 85–89, 91–95, 97,
    99–101, 103, 109, 110, 116,
    127–129, 158

T

Taylor's method, 56, 61, 62, 74
transfer
    function, 88– 91, 94, 95, 97,
        99–103, 109, 110, 113,
        115–119, 121, 123–125,
        135, 142, 146, 149–151,
        164, 174, 179–181
    matrix, 91, 95, 97, 98, 100,
        110, 118, 123
tricycle, 62–64, 75–77, 79

V, W

vector field, 47–49, 56–61, 67,
    68, 73, 193, 194
wiring system, 98, 99, 101,
    102, 103, 115, 119, 120,
    135, 146, 151, 167, 174,
    199, 223

Other titles from

in

Control, Systems and Industrial Engineering

2014

DAVIM J. Paulo
Machinability of Advanced Materials

ESTAMPE Dominique
Supply Chain Performance and Evaluation Models

FAVRE Bernard
Introduction to Sustainable Transports

MICOUIN Patrice
Model Based Systems Engineering: Fundamentals and Methods

MILLOT Patrick
Designing Human–Machine Cooperation Systems

MILLOT Patrick
Risk Management in Life-Critical Systems

NI Zhenjiang, PACORET Céline, BENOSMAN Ryad, REGNIER Stéphane
Haptic Feedback Teleoperation of Optical Tweezers

OUSTALOUP Alain
Diversity and Non-integer Differentiation for System Dynamics

REZG Nidhal, DELLAGI Sofien, KHATAD Abdelhakim
Joint Optimization of Maintenance and Production Policies

STEFANOIU Dan, BORNE Pierre, POPESCU Dumitru, FILIP Florin Gh., EL KAMEL Abdelkader
Optimization in Engineering Sciences: Metaheuristics, Stochastic Methods and Decision Support

2013

ALAZARD Daniel
Reverse Engineering in Control Design

ARIOUI Hichem, NEHAOUA Lamri
Driving Simulation

CHADLI Mohammed, COPPIER Hervé
Command-control for Real-time Systems

DAAFOUZ Jamal, TARBOURIECH Sophie, SIGALOTTI Mario
Hybrid Systems with Constraints

FEYEL Philippe
Loop-shaping Robust Control

FLAUS Jean-Marie
Risk Analysis: Socio-technical and Industrial Systems

FRIBOURG Laurent, SOULAT Romain
Control of Switching Systems by Invariance Analysis: Application to Power Electronics

GRUNN Emmanuel, PHAM Anh Tuan
Modeling of Complex Systems: Application to Aeronautical Dynamics

HABIB Maki K., DAVIM J. Paulo
Interdisciplinary Mechatronics: Engineering Science and Research Development

HAMMADI Slim, KSOURI Mekki
Multimodal Transport Systems

JARBOUI Bassem, SIARRY Patrick, TEGHEM Jacques
Metaheuristics for Production Scheduling

KIRILLOV Oleg N., PELINOVSKY Dmitry E.
Nonlinear Physical Systems

LE Vu Tuan Hieu, STOICA Cristina, ALAMO Teodoro, CAMACHO Eduardo F., DUMUR Didier
Zonotopes: From Guaranteed State-estimation to Control

MACHADO Carolina, DAVIM J. Paulo
Management and Engineering Innovation

MORANA Joëlle
Sustainable Supply Chain Management

SANDOU Guillaume
Metaheuristic Optimization for the Design of Automatic Control Laws

STOICAN Florin, OLARU Sorin
Set-theoretic Fault Detection in Multisensor Systems

2012

AÏT-KADI Daoud, CHOUINARD Marc, MARCOTTE Suzanne, RIOPEL Diane
Sustainable Reverse Logistics Network: Engineering and Management

BORNE Pierre, POPESCU Dumitru, FILIP Florin G., STEFANOIU Dan
Optimization in Engineering Sciences: Exact Methods

CHADLI Mohammed, BORNE Pierre
Multiple Models Approach in Automation: Takagi-Sugeno Fuzzy Systems

DAVIM J. Paulo
Lasers in Manufacturing

DECLERCK Philippe
Discrete Event Systems in Dioid Algebra and Conventional Algebra

DOUMIATI Moustapha, CHARARA Ali, VICTORINO Alessandro, LECHNER Daniel
Vehicle Dynamics Estimation using Kalman Filtering: Experimental Validation

HAMMADI Slim, KSOURI Mekki
Advanced Mobility and Transport Engineering

MAILLARD Pierre
Competitive Quality Strategies

MATTA Nada, VANDENBOOMGAERDE Yves, ARLAT Jean
Supervision and Safety of Complex Systems

POLER Raul *et al.*
Intelligent Non-hierarchical Manufacturing Networks

YALAOUI Alice, CHEHADE Hicham, YALAOUI Farouk, AMODEO Lionel
Optimization of Logistics

ZELM Martin *et al.*
I-EASA12

2011

CANTOT Pascal, LUZEAUX Dominique
Simulation and Modeling of Systems of Systems

DAVIM J. Paulo
Mechatronics

DAVIM J. Paulo
Wood Machining

KOLSKI Christophe
Human-computer Interactions in Transport

LUZEAUX Dominique, RUAULT Jean-René, WIPPLER Jean-Luc
Complex Systems and Systems of Systems Engineering

ZELM Martin, *et al.*
Enterprise Interoperability: IWEI2011 Proceedings

2010

BOTTA-GENOULAZ Valérie, CAMPAGNE Jean-Pierre, LLERENA Daniel, PELLEGRIN Claude
Supply Chain Performance / Collaboration, Alignement and Coordination

BOURLÈS Henri, GODFREY K.C. Kwan
Linear Systems

BOURRIÈRES Jean-Paul
Proceedings of CEISIE '09

DAVIM J. Paulo
Sustainable Manufacturing

GIORDANO Max, MATHIEU Luc, VILLENEUVE François
Product Life-Cycle Management / Geometric Variations

LUZEAUX Dominique, RUAULT Jean-René
Systems of Systems

VILLENEUVE François, MATHIEU Luc
Geometric Tolerancing of Products

2009

DIAZ Michel
Petri Nets / Fundamental Models, Verification and Applications

OZEL Tugrul, DAVIM J. Paulo
Intelligent Machining

2008

ARTIGUES Christian, DEMASSEY Sophie, NÉRON Emmanuel
Resources–Constrained Project Scheduling

BILLAUT Jean-Charles, MOUKRIM Aziz, SANLAVILLE Eric
Flexibility and Robustness in Scheduling

DOCHAIN Denis
Bioprocess Control

LOPEZ Pierre, ROUBELLAT François
Production Scheduling

THIERRY Caroline, THOMAS André, BEL Gérard
Supply Chain Simulation and Management

2007

DE LARMINAT Philippe
Analysis and Control of Linear Systems

LAMNABHI Françoise *et al.*
Taming Heterogeneity and Complexity of Embedded Control

LIMNIOS Nikolaos
Fault Trees

2006

NAJIM Kaddour
Control of Continuous Linear Systems

Lightning Source UK Ltd.
Milton Keynes UK
UKHW011834090221
378516UK00003B/178